MAN AND BIRDS

This book is one of a series, *Pegasus Topics in Biological Science,* prepared under the auspices of the Biological Sciences Curriculum Study.

MAN AND BIRDS

Evolution and Behavior

***

Andrew J. Meyerriecks

A Biological Sciences Curriculum Study Book

Pegasus

A DIVISION OF

The Bobbs-Merrill Company, Inc., Publishers

INDIANAPOLIS NEW YORK

Printed in the United States of America
Library of Congress Catalogue number 79-175222
ISBN 0-672-63558-5 (pbk)
Second Printing

Acknowledgments

The author, BSCS, and the publisher are grateful to the following for permission to reprint.

Figure 1: Taken from *The Pigeon,* 1957, by permission of the author, Wendell M. Levi. *Figure 2:* Courtesy of Robert Meyerriecks. *Figure 3:* Courtesy of British Ornithologists' Union. *Figure 4:* Courtesy of Robert Meyerriecks. *Figure 5:* Roger Tory Peterson. *Field Guide to Western Birds.* By permission of Houghton Mifflin Company. *Figure 8:* Photo by the author. *Figure 9:* Courtesy of Robert Meyerriecks. *Figure 10:* Courtesy of Robert Meyerriecks. *Figure 11:* Illustration by Robert M. Mengel, reproduced by permission of Yale University Press from *Handbook of North American Birds,* Vol. I. Ed. by Ralph S. Palmer, p. 485. Copyright © 1962 by Yale University. *Figure 12:* Photo by the author. *Figure 13:* Illustration by Robert M. Mengel, reproduced by permission of Yale University Press from *Handbook of North American Birds,* Vol. I. Ed. by Ralph S. Palmer, p. 450. Copyright © 1962 by Yale University. *Figure 14:* Redrawn from *Darwin's Finches* by David Lack. Cambridge University Press. *Figure 15:* Courtesy of Biological Sciences Curriculum Study. *Figure 16:* From Bowman, R. I.: Evolutionary patterns in Darwin's finches. California Academy of Sciences Occasional Paper no. 44, 1963. *Figure 17:* From David Lack's, *Darwin's Finches.* Cambridge University Press. *Figure 18:* Photo provided by the author. *Figure 19:* Redrawn from *The Auk*, Vol. 73, 1956, by permission American Ornithologists' Union. *Figure 20:* From Proceedings, Linnaean Society of New York, number 50–51. *Figure 21:* From a drawing by George Miksch Sutton in *Fundamentals of Ornithology* by Van Tyne and Berger. *Figure 22:* From *Island Life* by Sherman Carlquist. Doubleday and Company, Inc. *Figure 23:* From a drawing by George Miksch Sutton in *Fundamentals of Ornithology* by Van Tyne and Berger. *Figure 24:* G. Ledyard Stebbins, *Processes of Organic Evolution* © 1966. Reprinted by permission of Prentice-Hall, Inc. *Figure 25:* Redrawn from *Life of the Robin* by David Lack with permission of H. F. and G. Witherby, Ltd. *Figure 26:* Redrawn from *Life of the Robin* by David Lack with permission of H. F. and G. Witherby, Ltd. *Figure 27:* Redrawn from *Life of the Robin* by David Lack with permission of H. F. and G. Witherby, Ltd. *Figure 29:* Courtesy of Robert Meyerriecks. *Figure 31:* From a drawing by George Miksch Sutton in *Fundamentals of Ornithology* by Van Tyne and Berger. *Figure 32:* Redrawn from material originally published by the University of California Press; reprinted by permission of The Regents of the University of California. *Figure 33:* Redrawn from *Life of the Robin* by David Lack with permission of H. F. and G. Witherby, Ltd. *Figure 35:* Redrawn from the *Herring Gull's World by* Niko Tinbergen (Collins, London, 1953). *Figure 36:* Redrawn from *The Auk*, Vol. 73, 1956, by permission American Ornithologists' Union. *Figure 38:* Courtesy of Robert Meyerriecks. *Figure 43: Scientific American,* Dec. 1958. *Figure 44: Natural History,* June-July 1962. *Figure 47:* Redrawn from a drawing by Don R. Eckleberry in *Life Histories of Central American Birds* by Alexander F. Skutch, 1954. *Figure 48:* From a drawing by George Miksch Sutton in *Fundamentals of Ornithology* by Van Tyne and Berger.

Contents

Editor's Preface

City or town, street or meadow, telephone pole or maybe tree—wherever man is, so are birds. Birds are even in places where man usually isn't. This commonplaceness, coupled with their coloration and calls, and maybe even a tinge of jealousy of their ability to fly away from it all, accounts for the large measure of interest man shows in birds. Birds are honored as national symbols, including that of the United States, and in some of the world's great myths and romantic tales. Most any book dealing with birds thus finds a ready welcome among professionals and amateurs alike. Andrew Meyerriecks's book will be no exception.

However, this is no ordinary or typical "bird book." In the hands of a knowledgeable scientist and skillful writer, two most intriguing aspects of modern biology, evolution and behavior, are woven together with threads of what birds do. Birds behave: they feed and excrete, fly and roost, court and mate, nest and brood, and they communicate visually and aurally. In a word, they respond to changes in their environment, changes that are abrupt like the crack of a gun or the appearance of a predator, and changes that are subtle like the gradual lengthening of daylight or the enlarging of their population. The capacity to behave, to respond to these environmental changes is the outcome of evolutionary process. The failure to respond or to be incapable of responding to change is also an evolutionary outcome, but one which leads to individual death and species extinction.

This is what Dr. Meyerriecks's book is about, not behavior nor evolution as separate topics, but as one and using birds as the means for doing so. The fascinating thing, I believe you will find, is how much and how easily you will learn of both—and of birds. For those who are interested in the topic of evolution, you will find another book in this series of interest, Gastonguay's *Evolution for Everyone*. Also, some of the chemical basis of behavior is explored in LeBaron's *Hormones: A Delicate Balance*.

Edward J. Kormondy
Olympia, Washington

Part One

*

MAN AND BIRDS
Evolution

I

Some Darwinism

Variation and Artificial Selection

Charles Darwin put the stamp of evolution firmly and resoundingly on the world of biology with the publication of his now famous book *On the Origin of Species* in 1859. It is true that Darwin's various theories were attacked repeatedly during the decades following publication of his book; it is true that Darwin's mechanism of inheritance has been completely discarded in the light of Gregor Mendel's discoveries and the evidence from more recent researches in the field of genetics; but it is also true that Darwinian evolution remains as the soundest building block in the edifice that we call "modern biology."

The first two chapters of the *Origin* (as it is usually referred to) are devoted to an examination of *variation*. Darwin realized from personal observations that all of the individuals making up a species or kind of living thing are not identical. They may vary widely in a large number of attributes: size, color, pattern, longevity, fecundity, habits—to name but a few sources of variability. Darwin also observed that man took advantage of such widespread variation found among the individuals that made up the different plants and animals he had brought under domestication. Plant and animal breeders selected those individuals that showed characteristics desired by man and then allowed them to mate. It was irrelevant to man whether the attributes he selected were of any benefit to the organisms being bred. Such a process is called *artificial selection* and can be viewed as follows:

Fact: there is variation, much of it with an inherited basis
Fact: there is deliberate selection by man
this leads to . . .
Consequence: some variants are saved
Consequence: some variants are eliminated
bringing about . . .
Fact: there are some variants that are saved
Fact: there is continued artificial selection
which finally leads to . . .
Consequence: there is an accumulation of variant characteristics which equals the development of a *new breed or variety* of domestic plant or animal.

Let's look at these facts and consequences more closely by examining how man has artificially selected from the variations available in his stocks of domestic birds. The first fact—variation with inheritance—should be obvious to anyone who has ever raised or even casually examined domestic chickens, ducks or pigeons. Darwin used the domestic pigeon as a prime example of what artificial selection can accomplish. In fact, Darwin carefully questioned pigeon breeders about their techniques and results; he joined several pigeon fanciers' clubs; and he kept and bred many varieties of pigeons from his native England and elsewhere in Europe. Good scientist that he was, Darwin kept detailed and voluminous records of the activities of the pigeon breeders (including himself), but his humanity is clearly revealed in a letter he wrote to W. D. Fox in which he admits that pigeon breeding was a horrible bore!

Most aspects of the color, pattern and plumage of the domestic pigeon have been subject to artificial selection by pigeon fanciers. The pouter, carrier, fantail and tumbler are but a few of the many breeds or varieties of domestic pigeons developed solely through man's intervention. (See Figure 1.) But what was the nature of the original ancestor of these myriad kinds of domestic pigeons? Where and when did artificial selection by man begin?

It is definitely known that the ancestor of the domestic pigeon is the wild Rock Pigeon, or Rock Dove, *Columba*

Figure 1. Varieties of domestic pigeons brought about by man's artificial selection. All are descendants of the wild Rock Pigeon (Columba livia).

livia. This species is an inhabitant of cliffs and caves and similar remote areas from western Europe to western Asia, south throughout the Mediterranean region and north Africa, as well as India. The Rock Pigeon is still a common wild bird in many parts of this vast range. Where and when domestication first took place is not definitely known. Recent research puts it perhaps as long ago as 4500 BC in Mesopotamia. Furthermore, white varieties of domestic pigeons were well established during the heyday of the Greek and Roman empires.

Man selected from the natural variability present in the descendants of the wild Rock Pigeon; he was able to demonstrate the inherited basis of this variability by simply breeding together the appropriate birds and examining their offspring for the desired traits. He selected those offspring with the characteristics closest to man's wants, bred them, and thus continued the desired trait generation after generation, or at least as long as man desired. Hence, consequences one and two result. Man saves those variants of interest or utility to himself, but he discards the unwanted ones. Further selection by man intensified the desired attributes, whether this was whiteness of plumage, fullness of tail, peculiarity of flight, or whatever; the development of a new breed or variety was thus assured. Such a process of artificial selection has been going on for thousands of years and continues today all over the world in hamlets and scientific laboratories, by amateurs and professionals.

Darwin struggled constantly to come up with a satisfactory mechanism of inheritance; that is, he tried to explain how new variations arose or were produced and how they were transmitted from parent to offspring. In fact, he published in 1868 a work entitled *The Variation of Animals and Plants Under Domestication* in which he labored to explain the inherited nature of variability. All in vain. Completely unknown to Darwin, an Augustinian monk working in a monastery garden in what is now Czechoslovakia had published the results of his studies on heredity two years before the appearance of Darwin's book. The monk was Gregor Mendel, and his work consisted of a series of carefully planned and executed studies of inheritance in the common garden pea, *Pisum sativum*. Mendel's researches were detailed, quantitative and well analyzed; he had found the answer to one of Darwin's most vexing problems. Darwin finally resorted to a scheme of "blending" inheritance, now known to be completely erroneous, to explain variation and its transmission to subsequent generations. By contrast, Mendel clearly demonstrated that inheritance was based on the passage of discrete, nonblending "elements" or "factors" (to use Mendel's terminology) from one generation to the next. Today the factors of Mendel are called genes, and we know full well that an understanding of variation requires a knowledge of genes that embraces their chemical structure, their interactions with one another, their activities during development, and how alterations in their structure may provide new sources of variability.

Prior to the rise of genetics as a science (about 1900), artificial selection was imprecise, although it was remarkably successful in the long run. It was a slow process based largely on trial-and-error techniques. After the advent of heredity as a science in its own right, selection was based on a sound knowledge of the genetic makeup of the organisms being studied from the viewpoint of the breeder of new kinds of domestic plants and animals. How did their genes act and interact? What crosses or matings should be made in order to achieve the desired results? Problems of quality and yield could now be tackled from a firmer base; the probable variation found in the progeny from a cross could be predicted from an analysis of the variation in the parents. All well and good. However, we are still faced with the same problem that confronted Darwin over one hundred years ago.

Variation and Natural Selection

The problem is this: if *man* does the selecting of characteristics with his domesticated plants and animals, who or what does the selecting in *wild* populations?

In his autobiography Darwin wrote:

> I soon perceived that selection was the keystone of man's success in making useful races of animals and plants. But how selection could be applied to organisms living in a state of nature remained for some time a mystery to me.

The mystery was resolved in October, 1838, when Darwin read the famous *An Essay on Population* by Malthus. Based in part on his reading of Malthus and his own observations, Darwin later wrote:

> . . . being well prepared to appreciate the struggle for existence which everywhere goes on from long-continued observation of the habits of animals and plants, it at once struck me that under these circumstances favourable variations would tend to be preserved, and unfavourable ones to be destroyed. The result of this would be the formation of new species.

In other words, under the conditions of competition in which plants and animals live, those variations would be preserved that tended to increase the organism's ability to leave fertile, adapted offspring, while those variations that tended to decrease this ability would be eliminated. Thus, in a state of nature, selection would act automatically, which is why Darwin called the process *natural selection.*

Using the analogy of artificial selection, Darwin could argue as follows:

Fact: there is a rapid increase in the numbers of a species, but . . .
Fact: the total numbers of that species remain "constant"
this leads to . . .
Consequence: there is a "struggle for existence"
bringing about . . .
Fact: there is a "struggle for existence"
Fact: there is variation with inheritance
this leads to . . .
Consequence: there is a "survival of the fittest"
and finally . . .
Fact: there is a "survival of the fittest"
Fact: the environment changes
ending with . . .
Consequence: descent with modification = *evolution*.

Reproductive Potential

Let's examine these facts and consequences as we did with those concerned with artificial selection. The first fact—a rapid increase in numbers—is a fundamental characteristic of life. Many organisms have a fantastic reproductive or biotic potential. Female fish of a number of species may produce millions of eggs during the course of a single breeding season. The adult stages of some internal parasites are virtual reproductive machines, producing streams of eggs vital for the survival of the species. Even on a lesser scale reproductive accomplishments are impressive. The females of some frog and toad species may lay thousands of eggs in clusters or string-like groupings. Quail, grouse and other relatives of the domestic chicken may incubate clutches of a dozen or more eggs.

Charles Darwin deliberately selected the elephant as an example of an extremely slow-breeding organism, yet he was able to calculate that one pair of elephants could have millions of descendants in less than 800 years if all that were born lived to reproduce in turn. The English Starling, *Sturnus vulgaris,* and the English, or House, Sparrow, *Passer domesticus,* were both introduced into the United States from Europe with startling results. Both species spread from coast to coast rapidly, and now they are common breeding birds in most parts of the country even though both species started this dramatic increase in geographic range from small populations deliberately introduced by man on the East coast.

A more recent, equally dramatic range expansion and increase in population size is shown by the Cattle Egret, *Bubulcus ibis,* but this example differs from the starling and sparrow in that the Cattle Egret was not introduced by man. (See Figure 2.) This relative of the herons and bitterns is a native of the Old World, and the evidence now available suggests that African egrets made it across the Atlantic Ocean to South America on their own power. Cattle Egrets are strong fliers, are a migratory species, and they are highly social; furthermore, the distance from western Africa to the northeastern part of South America is not that great for an egret-sized bird to conquer. In addition, a close relative of the Cattle Egret is known to have flown a similar route in the past few years. A Little Egret, *Egretta garzetta,* banded in a heronry in Spain, was recovered on the island of Trinidad off the coast of South America a few months later.

After its initial landfall in South America the Cattle Egret began to spread in that continent and north towards

Figure 2. The Cattle Egret (Bubulcus ibis), *shown with a typical feeding associate, crossed the Atlantic and is now flourishing in the New World.*

Central and North America as well as the isles of the Caribbean. In 1953 Samuel Grimes of Jacksonville, Florida, found *one* nest of the Cattle Egret on an island in Lake Okeechobee, Florida. Today, colonies of this species numbering in the thousands can be found in Florida and elsewhere, and the bird has spread throughout much of the United States and even into Canada. Westward, Cattle Egrets have been observed in the Hawaiian Islands as well as other places in the wide Pacific Ocean—virtually a circumnavigation of the globe in far less than a century! Each year brings reports of new sight records, new nesting occurrences and new population buildups, repeating essentially the history of the English Starling and the House Sparrow with respect to range and population expansion. This is true reproductive potential, the kind discussed by Darwin. But the earth is not covered with Cattle Egrets or houseflies or any other species capable of remarkably high reproductive rates. Why?

Limits to Population Growth

This brings us to Darwin's second fact: total numbers of a species remain "constant." Suppose we were to make a census of robins each spring in our own backyard. The American Robin, *Turdus migratorius,* is a common breeding bird over much of the United States and is familiar to most American citizens. It is an easily observed species, and it has a tendency to nest near human dwellings, which makes our census easier to conduct. We might even make our study more exact by having a bird-bander friend capture our backyard adult robins unharmed with the aid of a trapping device called a "mist net." Our bander friend could then place tiny colored bands or rings around the legs of each robin so that we could tell each individual apart. Thus we could record who mated with whom, how many young were produced by each pair, whether or not the same robins returned to our yard each year, whether re-matings took place, and so forth. Robins may rear two or perhaps three broods each year, so that by the end of each summer our backyard might contain a large number of robins, both adults and their progeny of the breeding season. It might even look as though robins were about to inundate the area; however, when we made our annual robin census the following spring, before actual nesting began, we would find that about the same number of breeding adults had returned to our area to nest as in previous seasons. Our censuses taken over a long period would reveal that the number of breeding robins tended to remain about the same year after year. What happens to the annual surplus?

The answer is that something is limiting population growth; something is preventing the robins from reaching their reproductive or biotic potential. Sometimes the total of factors tending to limit population size is lumped together under the heading of *environmental resistance.* An individual may be eliminated purely by chance—a lightning strike or a rock dislodged from a cliff. Severe storms, floods and droughts may wreak havoc. A dramatic example was provided a few years ago in East Africa during a prolonged drought. Thousands of young flamingos were eliminated because of heavy encrustations of alkali mud on their feet and bills. The unfortunate birds could not move about, nor could they feed or be fed by their parents. Disease may strike an individual here or there, or epidemics may wipe out entire populations. The Passenger Pigeon, *Ectopistes migratorius,* was at one time superabundant in the United States; the ornithologist Alexander Wilson once saw a flock he estimated at 2,000 million birds. Now this species is extinct. There is speculation that unrestricted hunting, especially during the critical breeding season, may have reduced the populations of this bird to a low level where some disease of epidemic proportions gave it the *coup de grace.* Domestic pigeons and many of their close wild relatives harbor a parasite that may cause a parasitic disease known as trichomoniasis; a particularly virulent strain of this parasite may spread rapidly through the flocks and decimate them. Since pigeons pump food into the mouths of their young and the parasites are found commonly in the upper part of the digestive tract of the bird, transmission of the parasite from adult to young is virtually assured. Perhaps a similar parasite—a virulent strain perhaps—was spread quickly from one passenger pigeon to another, and the death knell of the species was sounded.

It is clear that predators will take individuals from a

population, sometimes in great numbers; thus, the size of the prey species' population will be kept in check. I once saw a large number of Ring-billed Gulls, *Larus delawarensis,* descend on a huge, milling throng of fiddler crabs on a mud flat in Florida. The predatory gulls gorged themselves to the point where some of them had difficulty in taking flight. Some species of predatory birds have evolved a specialized "food-finding call" to alert fellow members of the species to the presence of a superabundant source of food. Hubert Frings and his associates discovered such a call in the Herring Gull, *Larus argentatus.* Tape recordings of the call were made, and when they were played back a large number of individuals of this species were attracted to the sound source. The English student of animal behavior W. H. Thorpe believes that such calls are produced only when the food source is temporarily superabundant. Thus, a large number of individuals of that species can gorge themselves until the food supply runs out or disappears. If food is in short supply, the finder keeps that fact to himself, thus insuring that he will get enough to sustain himself. If a food-finding call were to be given, there would not be enough to go around. My fiddler-crab observation in Florida fitted Thorpe's scheme nicely: the crabs were superabundant, but they dispersed into the surrounding mangroves in a short time; only then did the gulls leave to forage elsewhere.

Overcrowding itself may bring about reproductive adjustments within the organism. For example, if mice are deliberately overcrowded in a laboratory experiment, severe effects may follow: individual growth slows or is stopped; sexual maturity in young mice is delayed or even prevented; reproduction by the adults is reduced; behavioral abnormalities appear in the young; and so forth. Reduction of the population size restores a more normal situation. Overcrowding may bring about spectacular mass flights as in the migratory locust, *Locusta migratoria.* In this species there are two kinds, or phases, of locusts—the solitary ones and the migratory ones. The young solitary locusts are sedentary, lack wings, do not move about very much, and in general have a protective coloration. As the population of solitaries in an area builds up, more and more of the migratory (gregarious) phase types are produced. As adults their wings are longer, legs are shorter, and their coloration is far less protective in nature when compared with the solitary phase individuals. When the population is properly synchronized, a swarm of flying locusts erupts and may fly hundreds of miles, devastating man's crops along the way. The population of migratory, or gregarious, phase locusts crashes, but a few individuals survive. Solitary phase locusts are again produced, and the cycle begins anew.

Competition for food, for shelter, for a mate, for any resource in the environment, may bring about population adjustments. The social structure of a species may be self-limiting with respect to population size. For example, a number of animal species show territorial behavior wherein certain individuals of a species obtain an area for themselves and defend it against intrusion by members of their own kind. This territory may be used for reproductive purposes; if so, exclusion of certain individuals will prevent them from breeding, even though they may be of reproductive age and in reproductive condition at the time of exclusion. This matter will be developed more fully later on in this book. In a harem society, like that found in such aquatic mammals as seals and sea lions, the young males must challenge and defeat the harem master in order to mate with the females.

Hence, a number of factors, singly or in combination, can act to limit potential population growth. The earth is not knee-deep in houseflies or egrets because a large number of environmental forces are acting to reduce or hinder catastrophic growth. If such growth does occur—gregarious locusts are but one example to many to be found in the animal kingdom—then eventually a population crash eliminates the excess. This leads us to Darwin's first consequence: there must be a "struggle for existence." Charles Darwin used the term "struggle for life"; he got the expression "struggle for existence" from one of his contemporaries, Alfred Russel Wallace. Both phrases suggest or imply combat, with victors and vanquished. They imply that fighting is rampant in the world of life. Actually, modern researches in the field of comparative ethology (essentially comparative animal behavior) indicate that most encounters between members of the same species are largely matters of bluff and threat. Actual physical contact and fighting leading to injury and even death are relatively rare. Darwin, however, did use the phrase "struggle for existence" metaphorically, but he did tend to over emphasize the competitive aspects of the phrase. Female codfish lay an incredible number of eggs; many of them are eliminated by chance, predators or the other factors we have discussed. This is a kind of "struggle for existence," but no combat whatever is involved. Animals do fight over food, but a number of mechanisms have evolved to get around the food shortage problem, none of which involves physical struggles over the food.

Avoiding the Struggle Over Food

For example, a number of animals store fat in their bodies during periods of food abundance; later they utilize the stored fat during periods of food scarcity. Many mammals such as woodchucks, hamsters and the like eat voraciously in late summer and early fall in the northern hemisphere;

then they hibernate during the long, cold, food-scarce winter, relying on their fat reserves to sustain them. Some animals cache or store food supplies when they are readily available, and then they use their stored food during periods of scarcity. One species, the European Nutcracker, *Nucifraga caryocatactes,* is dependent upon its food caches in order to rear its young properly. (See Figure 3.) These birds pair for life apparently, and the mated pair live on the same breeding territory in a spruce forest for the remainder of their lives. During the fall months, the nutcrackers are spending most of the daylight hours flying from their spruce forest territory to the nearest hazelnut woods. Here they gather hazelnuts, store them temporarily in their throats, then carry the load back to the territory in the spruce forest. Upon returning to their territory, the nutcrackers excavate shallow holes in the ground with their bills, dump a small pile of hazelnuts into the pit, then cover up the stored food. Sometimes they simply dump a

Figure 3. The European Nutcracker (Nucifraga caryocatactes) *gathers hazelnuts and stores them in shallow holes that it excavates on its spruce forest territory. The caches of nuts are utilized during the winter.*

small pile of nuts on the ground and cover it with lichens. Soon the snows start to fall and the cached food is completely hidden from the nutcrackers' view, or that of potential cache-robbers. But the nutcrackers have an amazing ability to recover their caches of nuts, even from under a covering of 18 inches of snow! Moreover, they rarely make mistakes. Swanberg studied these birds in Sweden and found that 86 percent of the hidden nut stores were found by the birds; additionally, this researcher discovered that the birds rarely made the mistake of digging through the snow to get a supply of nuts that had already been utilized. The bulk of the food fed to the young in late winter comes from the hazelnut supply stored by the parents of the previous autumn, so it is easy to see how vital this behavior is to the perpetuation of the nutcracker species.

Many birds migrate. That is, they make regular, annual movements to and from a breeding and wintering range; this behavior can function to avoid times of food shortage —leave an area when food supplies are dropping, return to it when they are being renewed. Another way to avoid times of low food supply is to become a nomad such as the Rosy Pastor, *Sturnus roseus*, a relative of the English Starling. Crossbills (genus *Loxia*) are also good examples of nomadic birds. Such species are opportunists, taking advantage of a temporarily abundant food supply and then leaving that area and wandering about until another such supply is found. When this has been used, they move on again. Some nomadic finch species found in Australia wander about their arid habitat until the brief rains begin to fall. Grasses in such regions quickly come to flower and seed, thus providing the finches with food. One researcher discovered that the male finches began to court females within a few seconds after the first rains started to fall, and shortly afterwards, functional sperm were to be found in the testes of these males, indicating an extremely close synchrony between the timing of reproduction of the finches and of the grasses they needed.

Avoiding Your Enemies

How about predator-prey relationships? Surely this is a kind of "struggle for existence." It involves death for the prey and the physical capture of its next meal for the predator. (See Figure 4.) It is true that the prey can and does try to defend itself; that is, it does "struggle for life," and it makes use of any defensive weapons it may have—horns, antlers, spurs, claws, hooves and so forth. But there are many ways to defend yourself without recourse to actual combat or the use of "weapons." You can simply fly away, or run like a rabbit, outmaneuvering or outdistancing a would-be captor. Avoidance of detection can be brought about by remaining absolutely motionless or by closely resembling the general background coloration. Many such protectively colored organisms do not move until the source of danger almost steps upon them. Such protective or adaptive coloration is widespread in the animal world, and the nicety of the resemblance is at times incredible. Larks of desert areas, for example, are beautifully sand-colored or resemble the rocks of their overall habitat. Where soils or substrates of different colors come into contact one can find several types of larks, each matching a different background coloration. The birds are reluctant to fly from one color background to another, as one German ornithologist discovered. Ground-nesting nighthawks and whip-poor-wills look like the forest debris on which they lay their eggs.

Many birds by contrast are very easy to see, and at first glance it would seem that they would be in difficulties with the predator problem; but one English biologist,

Figure 4. *The Louisiana Heron* (Hydranassa tricolor) *is an efficient predator of small fish and other aquatic organisms, seizing them scissor-fashion with its long, slender bill.*

Hugh Cott, was able to demonstrate that many such birds actually seem to flaunt their conspicuous coloration rather than attempt to hide it from potential enemies. Cott discovered that a large number of these birds were extremely unpleasant to the taste; their flaunting seems to be a kind of advertisement, a signal to would-be predators to leave the displayer alone for they would be wasting their time and energy to get a meal they could not eat because of its foul taste. Cott and his associates were able to make an edibility index, one that ranked birds according to the degree of their tastiness and the degree of protective coloration. Many of the protectively colored birds were in fact delicious, while the boldly colored and patterned species, the flaunters, were most displeasing to the taste. Cott formed a taste panel to test his theory; it included himself and several other humans as well as several kinds of mammals, and in general there was agreement on the rejection of the flesh of brightly colored birds and ready acceptance of protectively colored species as food items.

If you happen to be a "tasty species" there is still a way out besides looking like your background; the solution is in looking like another species which is in fact inedible. This phenomenon is known as mimicry, and there are many superb examples from the world of insects, but mimicry is little known and not well studied in birds. One interesting bird example, however, suggests that sometimes the predator can counter with his own brand of mimicry. The Zoned-tailed Hawk, *Buteo albonotatus,* occurs from the southwestern United States to Paraguay in South America. This bird of prey rather closely resembles the Turkey Vulture, *Cathartes aura,* which is also found throughout much of North, Central and South America. Potential prey animals are not much disturbed when they see small groups of vultures wheeling and gliding about in the sky above them because they will not be attacked by the inoffensive scavengers. But the Zoned-tailed Hawk takes advantage of this lack of response by flying with the vultures; the resemblance is increased because the hawk looks like a vulture. (See Figure 5.) When the mixed vulture-hawk group wheels over a prey species spotted by the hawk, the bird of prey suddenly swoops from the group of vultures and drops on its food in a swift act of aggression that belies its mimicry.

Turkey Vulture Zone-tailed Hawk

Figure 5. Rare example of bird mimicry is shown by Zone-tailed Hawk (Buteo albonotatus), *which resembles the Turkey Vulture* (Cathartes aura). *The hawk takes advantage of its resemblance to the vulture when it searches for prey.*

Avoiding Competition

Even if we do have direct competition for some environmental resource—say food—it is still possible to have evolved mechanisms that tend to reduce or even avoid competition. In other words, rather than competing for the same food you become a specialist better able to secure your needed food items as compared to your closest competitors living in the same habitat. My work on the foraging behavior of herons and egrets is instructive from this standpoint. On a turtle grass flat in Florida Bay it is possible to see as many as eight different species of herons and egrets foraging together, seemingly competing directly with one another since most of them are fish-eaters. (See Figure 6.) All are beautifully adapted for the pursuit, capture, handling and consumption of fish. All have long, slender legs, bills and necks which enable the birds to wade in water and strike rapidly out and down at their fish quarry. The prey is seized scissor-fashion between the tips of the mandibles. Actually, competition between these closely related and similarly adapted species is greatly reduced, as I've been able to show after watching the herons and egrets for many hours on the foraging grounds. Each species of fish-eaters is either a specialist in what it eats—it feeds on different kinds or species of fish—or else it feeds on different age classes of the same kind of prey. More important, however, is the way they go about foraging. Even the casual observer of herons in Florida Bay must be struck by the diversity of foraging techniques shown by the herons of this area. The Green Heron, smallest of the fish-eaters on the flats, has relatively short legs; thus, it tends to forage around the edges of the red mangroves, capturing tiny minnows in the shallows. (See Figure 7.) The much larger Great White and Great Blue Herons, by contrast, are to be seen out in the deeper

Figure 6. *Many heron species will feed in the same area, such as this turtle grass flat at the edge of the red mangroves in Florida Bay. They avoid competing by selecting different foods and by the way they go about searching for their prey.*

water, taking much larger kinds of fish. Reddish Egrets and Snowy Egrets, unlike the species just mentioned, do not stand still in the water and wait for their prey to draw near but actively seek them out, dashing and whirling about in rather shallow water and stirring their food into revealing themselves. (See Figure 8.) Another way to avoid competition is to work over the same foraging grounds but at a different time. This method is shown by the Black-crowned Night Heron and the Yellow-crowned Night Heron, both of which sleep in the red mangroves during the day but forage on the turtle grass flats at night; meanwhile their competitors have flown to roost. Life is diverse; there are numerous ways of avoiding struggle for the resources of the environment. Darwin's fact of a "struggle for existence," although true in part, must be considered warily because it is a catch phrase. As we have just seen, the "struggle" may be quite subtle. In view of our previous discussion of variation with inheritance, I think we can accept that fact as true at least for the moment. I'll be returning to this all-important matter of variation and its relationship to evolution and speciation.

Figure 7. *Green Herons* (Butorides virescens) *stand and wait in shallow water for their prey to draw within strike range.*

Figure 8. Whirling and dashing about, the Reddish Egret (Dichromanassa rufescens) *disturbs its prey, then stops and forms a canopy with its wings. It waits until a number of tiny fish gather in the shade of the canopy before it strikes.*

Who Is the Fittest?

This brings us to the next consequence, "survival of the fittest," another catch phrase, this time borrowed from Herbert Spencer. It is a good catch phrase, but it is also a loaded one; much misunderstanding has come from misuse of this phrase. Many think that strength and fierceness would tend to make an organism fitter than others; perhaps the image of a powerful predator such as the strong, fierce gray wolf comes to mind. But recent studies of wolf behavior have shown that this close relative of the domestic dog is a highly social mammal. Wolves typically hunt in packs, and they do not always succeed in bringing down their prey each time they start to forage. In fact, it has been shown recently that the larger prey of the wolf—such as moose—stand a good chance of escaping death if they stand their ground and challenge the pack. Even wolves go hungry at times.

The opposite of the strong, hard-running wolf is the timid little rabbit. It, too, can run rapidly and it is prolific. Most snakes are neither strong nor fierce; they may depend upon poisons, protective coloration and agility in order to escape. Tapeworms are not strong, fierce or fast, but they are incredibly prolific and successful in the living world. It would appear on brief examination that in the context of "survival of the fittest," the extremes would win out. Man clearly has this tendency to exaggerate the importance of the extremes: largest and smallest; strongest and weakest; fastest and slowest. In evolutionary matters, however, we find that it is the averages that seem to do best, to survive in the long run. Any departures from the norm—such as the lightest or heaviest—tend to be eliminated. Studies made of the behavior of Herring Gulls have shown that the most aggressive males in the colony are so belligerent that they don't get to mate with any of the females. Every time a female lands on the territory of an overly aggressive male she is threatened, attacked and driven from the area as if she were a rival male. Males that show more normal levels of hostile behavior succeed in reproducing. They, too, threaten and chase the females, but only early during the breeding season; their hostility quickly shifts to courtship and mating activities. Of course, human judgment as to what is fittest isn't worth too much: nature is the judge of what is fit and what is not fit. Furthermore, humans have a tendency to view but one aspect of an organism at a time: its speed, its size, and so on, but the only true test of survival, of fitness, is of course a test of the entire organism.

Man also tends to think of perfection and fitness as going hand in hand. But an organism that seems to be perfectly fit, that seems to be supremely adapted to its environment (in our judgment!), may actually be going down the blind alley of extinction. Such an organism becomes too dependent upon a rather narrow set of environmental factors. It is too specialized in its ways of living. It may be superior (fitter?) in securing a certain food resource when compared to other organisms, or it may be better at living in an isolated, special kind of habitat. But if its food supply drops below a critical level, or if its unusual habitat changes or disappears, then that specialist is in real trouble. The rats, the crows and the roaches live on; the specialist is but an evolutionary memory. What, then, do we mean by survival?

Survival is really a matter of success in leaving offspring; that is, survival is contributing to the next generation. (See Figure 9.) Of course, if an individual dies before reproducing, then it has not survived in an evolutionary sense because it has not contributed to the next generation. Here is a simple example; the white-tailed deer is one of the favorite prey of the gray wolf. A doe and its fawn are foraging in a forest and are attacked by a pack of wolves. The doe flees and its fawn falls victim to the hungry wolves. This doe has not survived in the evolutionary sense, even though it has escaped and lives to see another day. Its hereditary contribution to the next generation was lost with the death of its fawn. By contrast, if another doe stays and its fawn hides and escapes detection by the wolf pack, then this doe has survived in the evolutionary sense, even though devoured by the predators. Its hereditary contribution to the next generation rises from the grass

Figure 9. Evolutionary survival is a matter of producing offspring. Here an adult Black-crowned Night Heron (Nycticorax nycticorax), *on the right, feeds its young, insuring perpetuation of the species.*

after the wolves depart and continues its existence. Here, of course, we are dealing with biological or evolutionary survival. Each living individual manages to survive day by day—eating, sleeping, escaping—but if it does not reproduce, if it does not pass any hereditary material on to the next generation, then it has not survived for the "purposes" of evolution. The only real measure of fitness is survival, and the only measure of survival is a contribution to the next generation. What is the nature of this contribution?

What is passed from generation to generation is a packet or linear tape of genes, a code of information contained in the chromosomes found in the nuclei of the sperm and the egg. From this fertilized egg (or zygote) there develops an individual whose attributes are largely determined by the kinds of genes passed on to it from its parents. Some gene combinations will result in individuals well adapted to the environment in which they develop. Other gene combinations, by comparison, will render their carriers less well adapted, less fit to survive in the same environment. Bearers of the better-adapted gene combinations stand a better chance of reaching reproductive maturity and passing the "winning" combination on to their own offspring. Another way of viewing the matter is to consider natural selection as differential reproduction. The inferior gene combinations render their bearers less well adapted, and they make fewer contributions to the next generation simply because many do not reach reproductive age, or else cannot compete successfully with better-adapted individuals. The latter are more vigorous, more numerous, they survive to maturity and pass on these gene combinations to their own offspring. However, even the individuals with the winning combinations of genes are subject to difficulty if the environment changes. In this event, the now current, well-adapted gene combinations may not do as well as some other groupings of genes. This would lead to modification of those adaptations that enable the bearer to reach maturity and reproduce.

Fact six—changes in the environment—is easy to comprehend when we consider the geological history of the earth, abundantly illustrated for us in any modern textbook of geology. Uplift of mountains with subsequent erosion; vast areas of dry land once under the sea; marine fossils on the tops of mountains; these and a multitude of similar changes in earth history are testament to the fact that the environment is rarely stable. If it were, evolution would lead to groups of stable species, nicely adapted to the unchanging environment in which they lived and reproduced. There would be little diversity of life because of the overall sameness of the earth's crust. The changing nature of the environment now and in the past is clearly shown by the great diversity of life in the fossil record and all about us at present. The geological evidence is crystal clear: change is the norm of things.

We can now summarize our facts and consequences as follows, eliminating the catch phrases:

FACT		CONSEQUENCE
tendency to increase without limit		
+		
checks to population growth		
+		
inheritable variability	→	differential reproduction; that is, a selected portion of the population contributes to the next generation
differential reproduction	↙	
+		
changing environment	→	evolution

Species in Time and in Space

Before we see how the study of birds has contributed to our knowledge of evolution, we must understand the difference between Darwin's view of species and how they evolve as compared with recent, post-Darwinian ideas on the subject. Darwin essentially viewed species from the standpoint of time. Through differential contributions to the next generation and natural selection of the better-adapted individuals, there would be a gradual change in the nature of species. Darwin used the phrase "descent with modification." If we could go back to the starting point, to the origin, of a species in some time machine, and then follow the changing history of that species through time until the present (or until it became extinct), we would likely find a long series of small, gradual changes. Or, lacking a time machine, if we had a complete, unbroken series of fossils of this same species, we would just as likely find a long series of slight changes with the passage of time.

Descent with modification would be very difficult to establish along the line of fossils if we confined our scrutiny to *brief* time spans, but if we were to examine the beginning and end fossils, or else studied two series of fossils along the time span that were widely separated in time, we might be surprised to find striking differences. Indeed, the differences in these widespread time samples might be such that we would not hesitate to consider them as being representatives of two species of organisms. Descent with modification, brought about essentially by natural selection, would have led to a descendant species from an ancestral one.

However, Darwin called his great classic *On the Origin of Species*, but paradoxically he did not solve the problem of *how new species evolve!* He did demonstrate convincingly that species are modified by natural selection as time passes (descent with modification), and in the sense of "origin" we could consider that the species at the end of our time span has "originated" from the ancestral species as discussed above, but there is another way of viewing the origin of species. This approach to the species problem has occupied the attention of scores of biologists for many decades, and it is a way of viewing evolution to which the study of birds has contributed a great deal.

Suppose we enter a forested area and find that four different kinds of small birds inhabit it; we further suppose that all competent viewers of these four kinds of birds believe that they are very closely related, although representing four different species of birds. Now, if we were able to go backwards in time and trace the evolutionary history of these four species, we might conclude that we would travel along four separate time tracks until we reached the ancestors of each of the four species of forest birds. But Darwin and modern evolutionary biologists do not believe that this is so. Darwin and followers of his thinking believe that if we did go far enough back in time we would find *one common ancestor* for all four modern-day species of birds in our example. In other words, somewhere along the time span a *divergence* took place, perhaps several, which led to four separate evolutionary histories. Let's use the following diagram for further clarification, page 33.

Darwin viewed the origin of species as essentially a transformation of an ancestral form, by natural selection, which led with the passage of time to a descendant, new or different species. In the diagram, this can be viewed as tracing the history through time of one of the modern species A, B, C or D. They have descended with modification from the ancestral form. But, the diagram clearly indicates that species multiply in space as well as transform in

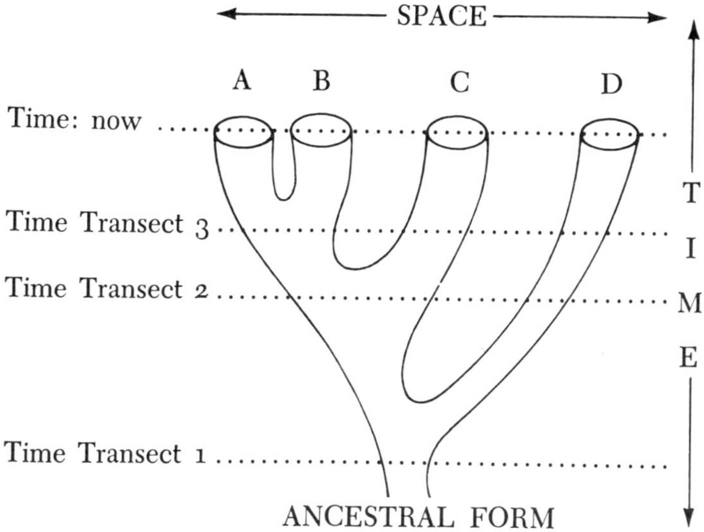

time. At time transect 1, perhaps thousands, tens of thousands, or even millions of years ago, there was only one species in the evolutionary history of our hypothetical group of forest birds. At this time transect the ancestral form may have been but slightly changed; perhaps the environment had not changed drastically or not enough time for major changes had passed. At time transect 2, by contrast, we find that some kind of a split, a divergence, had taken place. Now we find two descendant forms whereas previously we had but one. At time transect 3 we have three species resulting from two divergences. Finally, today we have four different species of forest birds, all descended from a single common ancestor. Species A, B, C and D all transform in time, but what brought about the divergences indicated in the diagram? We find that not only is there a transformation of species in time (Darwin's descent with modification), but there is also a multiplication of species in space. The latter process is known as speciation, and modern studies have contributed greatly to our understanding of this problem, one which vexed Darwin and one which is not completely solved today. As we shall see, Darwin's great contribution of natural selection as the prime guiding force of evolution, so to speak, is just as valid for speciation—the multiplication of species in space—as it is for the transformation of species in time.

II

Potential for Evolution in Bird Diversity

Individual Variation

Just as knowing about variation is the keystone to success in artificial selection of domestic stocks by man, a thorough knowledge of variation is necessary for an understanding of speciation in wild organisms. Since birds are the main subjects of this book, let's see what kinds of variation may be detected in these animals. Birds may show individual variation as they develop from the newly hatched chick to the final adult form. Simply compare a duckling Mallard with an adult of that species. In a host of birds the immatures are strikingly different when compared with the adults. This is true, for example, of the Little Blue Heron, *Florida caerulea.* The immature birds of this species are white while the adults are slaty-bluish on the body with a maroon neck. (See Figure 10.) When the immature Little Blue Heron is molting from the immature to the adult plumage, it presents a rather bizarre

Figure 10. *Adult and immature Little Blue Herons* (Florida caerulea) *are strikingly different in plumage coloration. The immature is pure white all over, while the adult (above) is dark plumaged.*

appearance, with dark feathers beginning to appear among the white ones. This pied pattern is very striking in the field. At times such bold differences between growth stages have fooled students of animal classification in the past. Some young hawks have been carefully described as one species while the adults were described and classified as yet another, until some observant naturalist discovered the mistake. In the United States in the last century it was believed that a Red-headed and a Brown-headed Woodpecker coexisted as two distinct species in the forests of the eastern part of the country. Again, careful field work by naturalists showed that the brown-headed "species" was really the immature stage of the red-headed one.

At times these developmental differences may be found, on careful examination, among the so-called adults. Hornbills are birds that have a large structure on the bill called a casque. In at least one species, found on the island of New Guinea, there are "young adults" and "old adults." Even though both age groups look alike, that is, they have similar plumage color and pattern, they can be told apart by the number of folds on the casque: 2-3 folds in young adults; 5-8 folds in old adults. In the Black-crowned Night Heron, *Nycticorax nycticorax,* which is virtually worldwide in its distribution in suitable aquatic habitats, the young or immature birds are quite different in color and plumage pattern when compared with the adults. Among the latter group, those breeding for the first time are often called "dusky" adults since they lack the immaculate white belly plumage of the final adult condition. An interesting evolutionary problem concerns a very closely related species, the Yellow-crowned Night Heron, *Nycticorax violaceus;* this species is widespread in North, Central and South America. Although the adults of these two night heron species are quite different in color and pattern of plumage, voice and other aspects of behavior, the immatures of the two species are very similar and are often confused by beginning bird watchers. (See Figure 11.) Later we shall examine some of the reasons for such similarities and differences between closely related species.

Seasonal Variation

The same individual may not only show great variation in plumage from nestling to immature to adult, but it may also show seasonal variation as an adult. The breeding adult male Scarlet Tanager, *Piranga olivacea,* of the New World is a vivid scarlet and black. In the fall, however, the male molts the brightly colored plumage and he assumes

Figure 11. *Adult (left) and juvenile (right) Yellow-crowned Night Herons* (Nycticorax violaceus) *have quite different plumages, but the juvenile Yellow-crown closely resembles the juvenile Black-crowned Night Heron. The adults of these close relatives are easily told apart.*

the yellow-green color characteristic of the females. Splendid Fairy Wrens, *Malurus splendens,* of Australia show similar seasonal changes, with the male having a brilliant plumage during the breeding season but one resembling that of the females during the nonbreeding season. Males of many ducks molt from a brilliant so-called "nuptial" plumage to a duller-colored one often referred to as the "eclipse" plumage. It is often one that resembles the plumage of the female, but there are exceptions. King Eider males (*Somateria spectabilis*) show a dramatic transformation in plumage color and pattern from eclipse to breeding season finery; if a beginner did not know the real relationship between the plumage types he would think they were two different species. All of us are familiar with those animals of the high arctic regions that are white in the winter and brown in the summer. This is an obvious adaptation to the overall background coloration as it changes seasonally, and such a pattern of coloration has clear survival value for the bearer. Ptarmigan, which are close relatives of birds such as grouse and prairie chickens, are good examples of birds showing clear-cut seasonal plumage changes. A bird need not molt its feathers in order to achieve color or pattern changes, however; feather wear can achieve dramatic results as in the color changes of the Snow Bunting, *Plectrophenax nivalis.*

Such seasonal changes in birds are not restricted to the feathers. In herons, for example, there are sharp differences in the coloration of the bill, the bare skin around and in front of the eyes, and the legs and feet, when these are viewed at different seasons of the year. For example, I studied the Green Heron, *Butorides virescens,* from this viewpoint. On its wintering grounds, say in the Keys of southern Florida, Green Herons are more or less nondescript; they have dull brownish or blackish bills, while the skin around the eyes tends towards a dull yellowish-gray. The legs and feet are a bright yellow or yellowish-orange. During northward migration, but especially upon arrival on the breeding grounds, their coloration is quite different. The bill of the heron is now a lustrous black; the skin around the eyes is a deep bluish-black, contrasting nicely with the brilliant yellow-orange eyes. But the legs and feet show perhaps the most dramatic changes: both become a glistening salmon pink. Both sexes show similar changes in coloration at the start of each breeding season, but the changes are more striking in the males. (See Figure 12.) Other heron and egret species show similar color changes in the body parts noted for the Green Heron, but no two species show the same combinations of colors. These changes at the start of the breeding season are clearly related to the courtship behavior of the male herons and egrets, a subject we shall explore in detail later on in the book.

Sometimes the sexes of a bird species are very difficult to tell apart by the usual method of observing them in the field and attempting to differentiate them on the basis of color, pattern, size, and so forth. Having them in the hand may not help either, so recourse to an examination of the primary sex differences (gonads, etc.) may be necessary for some purposes. Other than the primary differences, many species of birds show little variation in plumage color and pattern between the sexes. Other species, however, may show slight, moderate or even extreme sexual differences in outward appearance. These differences come under the heading of sexual dimorphism. One of the most famous cases in the bird world is that of the King Parrot, *Eclectus roratus,* of the New Guinea region in the southwest Pacific. In this species, the female is red and blue and has a black bill, while the male is green with an orange bill; both sexes were considered to be distinct

Figure 12. *On its breeding territory a male Green Heron shows brilliant color on its legs, feet, and bare skin around and in front of the eyes. Such colors serve to make the male more conspicuous as he displays. (Courtesy of Andreas Feininger.)*

species for almost one hundred years until the truth was determined. As we have previously seen, in the examples of the Scarlet Tanager and Fairy Wren, sexual dimorphism may be striking in the breeding season but virtually non-existent during the nonreproductive period. As with growth differences and individual variation, sexual differences in color, plumage and other attributes add to the store of variation to be analyzed by the student of bird evolution.

So far we have confined our examples largely to plumage—its color and pattern—but all attributes, all characteristics of an organism, can show variation. In birds, the length of the bill, wings, tarsus and tail may vary, as well as the weight, presence or absence of a crest, or the number of spots on the outer tail feathers. Much use of the variation in such characteristic features of birds has been made in the past, and analysis of similar variation goes on today. It is easy to see why, since color, pattern, length, and so forth, are easily observed and measured. But in recent years additional attributes have been subject to closer looks, especially since techniques for their detection and analysis have come along. Physiological, biochemical, cytological, ecological and behavioral characteristics, as well as others, have been added to the roster of attributes providing us with information on variation. Birds have not provided nearly as detailed information on blood group antigens and chromosomal types, but even here variability has been demonstrated.

Behavioral Variation

With the rise of comparative ethology, a comparative approach to problems in animal behavior, bird biologists now have at their disposal a tremendous amount of information on the variability of behavior. In their researches bird ethologists have concentrated on visual and auditory aspects of communication; this is natural, since birds and men are largely creatures of sights and sounds. With the introduction of the tape recorder and the parabolic reflector for recording and analysis of the taped sounds by means of spectrograms, biologists are now able to study variability in the songs and calls of birds; studies of song variation have given us new information on how bird species evolve. With respect to individual variation in songs and calls, birds are remarkably diverse. Some species show virtually no individual variability, while others demonstrate extreme variation in vocalizations. For example, a study of juncos made in the pine forests of Durango, Mexico, showed that of seventy individuals recorded, no two of them were precisely alike. In fact, some of the songs were so different that it was hard for the researchers to believe that they were being sung by birds of the same species.

Not only do individual birds in a population vary one from another, but recent work has shown the variability to be discovered in a single individual's songs. Many birds have several distinct utterances, but one Carolina Wren, *Thryothorus ludovicianus*, sang twenty-two different kinds of songs in twenty-four song outbursts! Other individuals of different species may have over one hundred kinds of variations to their basic song. So, variability in bird song may be at the level of the individual, the population, or—as I will develop fully later on—song may show trends in geographic variation as well.

Polymorphism

Another type of variability is known as "polymorphism," a term meaning many forms. In a number of animal species, including a variety of birds, we would be able to categorize individuals into two, three or more groups based on easily observed, rather sharp differences between them. We know that the individuals of a group may differ in length, weight, and so forth, but these are intergrading, often slight differences from one individual to another as compared to the polymorphic differences now being discussed. These are discrete, sharply separable differences. For example, the Reddish Egret, *Dichromanassa rufescens*, occurs on a number of islands in the Caribbean, in the Florida Keys, and along the gulf coast of Texas southward through Central America. Some individuals of this species are completely white; others are dark-plumaged. (See Figure 13.) This is not to be confused with the situation in the Little Blue Heron previously mentioned. Recall that in the Little Blue Heron the young birds are white, then develop a dark plumage through molt of their feathers. In the Reddish Egret, a bird is born either white or dark

Figure 13. *A striking kind of polymorphism is shown by the plumage of Reddish Egrets. Some individuals are dark colored, while others are pure white, regardless of age or sex.*

and does not change with age. This is a genetic difference that has nothing to do with either age or sex of the individual. If you go to a colony of breeding Reddish Egrets, say on Green Island off the Texas coast, you might be able to see dark males mated with dark females, or white males with white females, or birds of either sex mated with the opposite color type, or phase, as it is called. The Reef Heron, *Demigretta sacra*, widespread in the Pacific Ocean area, shows a similar type of plumage color polymorphism, but it may be complicated by the presence of a third kind of individual intermediate in coloration between the extremes of white and dark. Many of these polymorphic variations have been described in the past as distinct species until keen-eyed observers saw that they were color forms of one species. For example, the white phase, or type, of the Reddish Egret was once known as "Peale's Egret," and it was duly described and classified as a new species until a naturalist found them freely mating with the dark-phase birds. Polymorphism also affects attributes besides color: blood groups, presence or absence of spots and so forth.

Thus, all of the changes in growth, differences between the sexes, seasonal patterning of plumage, vocal repertoires, polymorphism—all of the characteristics we have been talking about provide an enormous storehouse of individual variability, a great supply of variation that can be acted upon by natural selection. Some of these variants will contribute better to the survival of their bearers than do those variants found in other individuals. The survivors in turn will pass on these favored or selected attributes to the next generation (or the potential for such attributes in the form of particular gene combinations); thus changes will be brought about with the passage of time. But these changes are taking place *within* populations. We find that in order for speciation to take place we must also have available for the working of natural selection some differences *between* populations. Let's turn now to a look at some examples of inter (between) population, or geographic, variation.

Geographic Variation

Those attributes such as size, color, plumage pattern, song, polymorphism, and so forth, that we looked at with respect to individual variability are also subject to geographic variation. For example, size is a commonly studied attribute from the standpoint of how it varies with geography. A cockatoo (kind of parrot) from the island of New Guinea showed geographical differences in the length of the wing when the populations making up this species were studied from one end of its range to the other. The length of the tail or its shape may vary from one part of the range to another. We may have, for example, one population of a species with long, narrow tails while another has short, rounded tails—the two populations being found in separate parts of the range. In some species we may find that a *character gradient* can be discerned, with a rather smooth progression from one extreme to another —say short tail in the south to long tail in the north, with every gradation in between. Such gradients are often referred to as *clines*, and they can be found in a number of attributes. At times a smooth cline may suddenly break, often where some climatic factor shows a sharp break itself. Then, on the other side of the climatic discontinuity, the cline may continue on a new smooth gradient.

Color and color pattern often show striking variation as we proceed from one end of the range of a species to another. On an archipelago in the Southwest Pacific, one species of a group of birds called "white-eyes" showed geographic variation in the color of the bill, forehead, belly and breast. In addition, this same species had populations with a large eye ring, a small eye ring or no eye ring at all. Sexual dimorphism, referred to earlier in our discussion of individual variation, may also show geographic differences. Again, an instructive example comes from a group of birds called "whistlers" from the Southwest Pacific. One species has males that are very brightly colored in one part of the range (the females are drab in coloration), while in another part of the range the males are "hen-feathered," that is, they closely resemble the female in plumage color and pattern. A small flycatcher from the same region of the Pacific Ocean shows a similar pattern, but adds a bit of a twist to its sexual geographic variation: on one island in their range it is the females who resemble the males!

Geographic variation in behavior is clearly shown by the song characteristics of those species of birds that occur widespread on a continent and also occur on some small, offshore islands. For example, the Blue Titmouse, *Parus caeruleus,* occurs in Great Britain and elsewhere in Europe, and it is also found in the Canary Islands off the western coast of Africa. A uniform type of song is sung by this species in Britain; furthermore, the song of the Blue Titmouse can be told quite easily from the songs of the other five species of titmice found in Britain. The latter are close relatives of the Blue Titmouse. On the island of Tenerife in the Canaries group, however, the Blue Titmouse shows a tremendous variability in its songs. Some of these songs resemble those of other titmouse species found in Britain, while others are quite different. It may be that this geographic variation in song types is because the Blue Titmouse is the only member of its group found on Tenerife. Throughout its range a bird species will be singing in competition, so to speak, with a number of other bird species, whether or not they are closely related. The sound environment may be quite complex, a situation that is more or less normal for continental environments. On a small, isolated island, by contrast, this same species may have no close relatives at hand, as in the Blue Titmouse of Tenerife, or at best one or perhaps two close relatives. Additionally, and very important for the study of song variation and geography, the sound environment will be much more simple on the island because there will be fewer species competing by singing. Sometimes geographic variation in song takes just the opposite turn: a species may have a rather complex song on the continent, but it "reverts" to a much simpler song on the island. These matters will be explored further a bit later in the book.

Another aspect of geographic variation has to do with the ecology of species; that is, their relationship with the physical and the living aspects of the total environment in which they live. All of us know of examples of woodland birds, such as woodpeckers or thrushes, or of desert birds (larks) or sea birds (petrels and albatrosses). But, if we examine a certain kind of habitat more closely we find some interesting relationships among the birds living therein. The titmice of Eurasia provide interesting examples (these birds are called chickadees in the United States). The Great Tit, *Parus major,* is the largest and most widely distributed titmouse of the Eurasian landmass. It shows interesting patterns of variability in its habitats throughout its vast range. In Japan, Korea and elsewhere in eastern Asia, this species lives in conifer forests and has a small, needle-like bill. Elsewhere, however, this species lives in broad-leafed woods and has a heavier, broader bill. The Great Tit may also inhabit quite arid country, as for example in central and southwestern Asia. By contrast, this same species avoids arid areas in its range in northwestern Africa. Structural modifications of the bill accompany such habitat differences.

In broad-leafed forests the Great Tit tends to have a broader, heavier bill than when it occurs in coniferous forests, for example, where its bill is smaller, slenderer, quite needle-like. A similar situation is found in the Blue Titmouse, previously mentioned with respect to geographical variation in song. This species occurs in gardens, parks and broad-leafed woodlands throughout a huge range in Eurasia. However, its habitat preferences are much narrowed toward the extreme northern and southern parts of its range in Europe. In northwest Africa, however, this species is once again broadly tolerant of a wide variety of habitats as it is in central Europe. Recall that the Blue Titmouse is the only member of its group present in the Canary Islands. In certain parts of this island group the tiny Blue Titmouse is found in mountain pine forests; yet, in other parts of its island range it avoids such habitats. It may be easy to quickly characterize a bird species as being "woodland," "desert," "aquatic," and so forth; but if we examine the ecology of that species more closely, we find more often than not that it shows a great deal of variation from place to place in its habitat needs. This kind of ecological variation, too, is subject to the action of natural selection.

Polymorphism you will recall is an especially clear-cut example of individual variability: one individual may be white while the other is dark, like the Reddish Egrets of the New World or the Reef Herons of the Pacific Ocean with their white, dark and intermediate forms. One human may be of blood type O, another A, while still a third may show blood type B or AB on being tested in the medical laboratory. Such polymorphic characteristics also show geographical variation. If you were to island-hop in the Caribbean region and count the frequency of the two-color phases of the Reddish Egret, you might conclude before your count that there would be equal numbers of both types in any population that you sampled. The truth is, however, that the frequency of the two forms varies from island to island, from region to region, seemingly in a haphazard or random manner. The same is true for the more complex phases of the Reef Heron. Why should one island have essentially all dark-phase birds, another island all white, and islands not too distant have varying ratios of the color types? What is the meaning of this interisland variability? What is the meaning of geographic variation? What relationship, if any, does it have to problems in speciation? Let's turn now to a consideration of birds on islands and birds on continents to pursue and attempt to answer these fundamental questions in evolutionary biology.

III

Isolation and Islands

The Galapagos Archipelago

> The natural history of these islands is eminently curious, and well deserves attention.

So wrote Charles Darwin about his visit to the Galapagos Islands, an archipelago that was to provide him and numerous later biologists with superb evidence for speciation, adaptive radiation and other evolutionary phenomena.

The Galapagos Islands (the name comes from a Spanish word meaning tortoise) lie on the Equator about 600 miles to the west of Ecuador in the eastern Pacific Ocean. The cold flow of the Humboldt Current sweeps north and west along the South American continent, out into the Pacific, and then passes by the isolated islands of Darwin. Cocos Island and Malpelo Rock are the nearest islands, and both are about 600 miles to the northeast. West of the Galapagos is the open Pacific for roughly three thousand miles.

The Galapagos Islands are owned by Ecuador, while Cocos Island belongs to Costa Rica. (See Figure 14.)

What did young Charles Darwin find on these remote islands that was to contribute so much to his ideas on organic change? As a young man, Darwin was appointed naturalist (with no pay) aboard HMS *Beagle*, an English ship under the command of Captain FitzRoy. The *Beagle* was to circumnavigate the globe and undertake a number of scientific investigations. The ship left England in 1831 and was not destined to return to its home shores until 1836. This voyage, the only one undertaken by Darwin in his lifetime, set the stage for all of his later efforts in evolutionary biology. It was truly one of the great sea voyages of all time. After detailed studies in and around South America, the *Beagle* headed for the Galapagos Islands and landed there in September, 1835. Then it was that the young naturalist discovered a flora and a fauna that was in

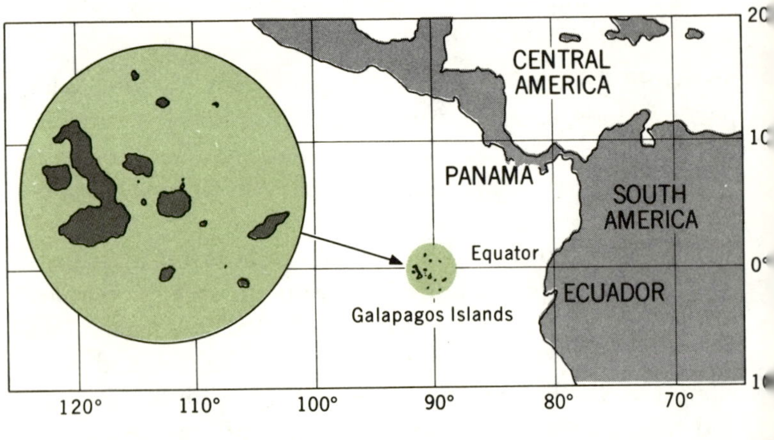

Figure 14. The Galapagos Islands, owned by Ecuador, are located about 600 miles to the west of that country in the eastern Pacific Ocean.

many respects very different from that of the mainland South America, which he had just left a short time before. And yet, the plants and animals of these remote islands were clearly related to those found on the distant continent. Darwin began to ponder about the similarities and the differences. Of the animals, Darwin saw much that startled him: huge lizards that swam in the sea and fed on marine plants; giant tortoises that showed remarkable variation from one island to another. But most importantly perhaps, he discovered a group of small, darkish-colored birds, obviously closely related, that showed striking diversity in the nature of their bills. There were kinds with large, heavy bills; others with thin, needle-like bills; some seemed to have bills that were intermediate between these two types. Furthermore, these birds varied from island to island in the archipelago. These strange, dull-colored birds that Darwin discovered on the Galapagos Islands are now called "Darwin's finches" in his honor. (See Figure 15.) Their detailed study has yielded many clues about the workings of evolution.

Darwin's Finches

What kind of birds are these dull-colored creatures that excited the mind of the young naturalist, as well as the minds of numerous evolutionary biologists ever since 1835? Darwin's finches are classified by some ornithologists in the family Fringillidae, the finch family. This large family includes such birds as buntings, grosbeaks, American sparrows, cardinals, crossbills, and so forth. In this arrangement, the group we call Darwin's finches are considered to be distinctive enough to warrant classification in their own subfamily—the Geospizinae. Other students of bird classification place these island finches in the family Em-

Figure 15. *On the Galapagos Islands, Charles Darwin discovered a group of dullish-colored birds—now called Darwin's finches in his honor—that showed great diversity of bill types. The study of these birds has revealed much about evolutionary processes.*

berizidae, and they include in this group the buntings, American sparrows, cardinals, New World honeycreepers, tanagers, and the like. In this scheme the Darwin's finches would be considered to constitute a tribe—the Geospizini. At any rate, the group consists of fourteen living species (there are no known fossils) distributed as follows: thirteen species of the finches occur only in the Galapagos Islands, while the fourteenth occurs only on (is "endemic" to) Cocos Island. Where did the Darwin's finches come from? What group of birds is ancestral to these inhabitants of the Galapagos? These and similar questions cannot be answered with any ring of authority at this time.

Bill Types Among the Finches

An examination of these dull-looking but fascinating birds reveals that in their structure and behavior, as well as in their habitats, they resemble birds found in South America such as seed-eating finches, insect-eating warblers and so forth. Much, but certainly not all, of this great diversity of feeding types among Darwin's finches is brought about by the great diversity to be found in the bills of these birds; in fact, much of the evolution of these birds has been concerned with features or adaptations connected with food-getting. Robert Bowman, a recent student of the group, has compared the shapes of the bills of the finches with the shapes of six kinds of pliers to indicate how the different bill types function. (See Figure 16.) Six species of the genus *Geospiza* have the heavy-duty, linesman's-pliers type of bill. These six species feed largely on seeds and some exposed insects. Although all six species feed extensively on seeds, it is important to note that the species differ in the kinds of seeds taken: large or small in size, soft or hard in texture. *Geospiza magnirostris* feeds

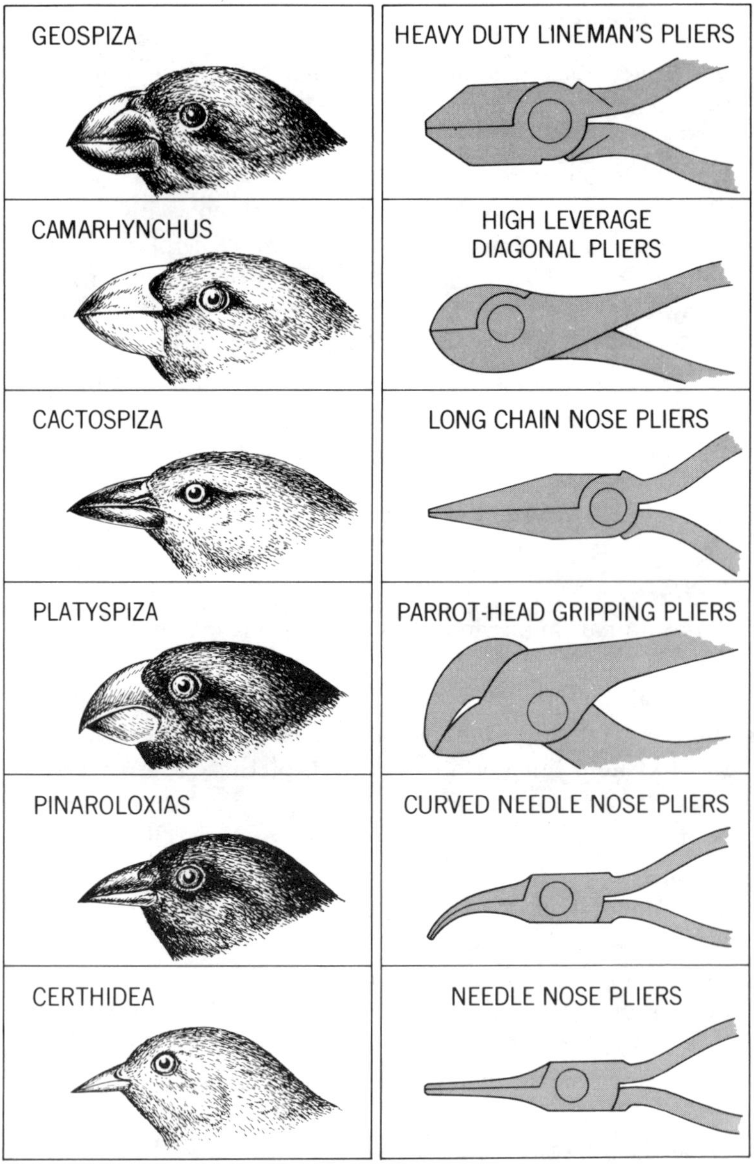

Figure 16. Robert Bowman, a student of Darwin's finches, has likened their bill shapes and functions to different kinds of pliers and how they work.

on small, hard seeds; *G. fortis* takes a variety of seeds with intermediate hardness; *G. fuliginosa* feeds on soft seeds of a wide variety. The nectar and fruits of a cactus commonly found in the Galapagos are taken by *G. scandens,* and this species also feeds on small seeds that are not overly hard. The feeding habits of *G. difficilis* and *G. conirostris* are not too well known, although both take seeds and perhaps insects and tissues of the cactus. With the bill type that they have, these six species of *Geospiza* are capable of crushing hard seeds between the bases of their powerful bills, and they are also capable of a strong bite at the tip of the bill.

Three species are classified in the genus *Camarhynchus* —*C. psittacula; C. pauper;* and *C. parvulus.* The bill and associated structures allow these three species to have quite strong biting at the tips of their bills and somewhat less powerful crushing between the bases of the bills than do the formerly described six species of *Geospiza.* In their diets the *Camarhynchus* species lean more towards insects and softer seeds. The insects are taken mainly from concealment by excavation of the wood by the finches. The bills of these species are described by Bowman as being of the high-leverage, diagonal-pliers type.

Soft fruits and insects largely make up the diet of the two species of the genus *Cactospiza*: *C. pallida* and *C. heliobates,* two most remarkable birds as we shall see momentarily. To use Bowman's pliers analogy once more, these two finches have the long chain, nose-pliers type of bill which permits powerful probing and strong tip-biting in their feeding ecology. The parrot-head, gripping type of pliers are used by Bowman to describe the bill of *Platyspiza crassirostris,* one of Darwin's finches that feeds on hard or soft seeds as well as fruits, buds, flowers and leaves. Strong crushing and tip-biting are the major adaptations of the bill of this bird. The weak, probing, grasping-tip type of bill of *Pinaroloxias inornata* enables this species —the only Darwin's finch to occur on Cocos Island—to feed on nectar, fruits and insects, although its overall biology is not well known. Finally, a needle-nosed-pliers kind of bill, adapted for strong probing and grasping, adapts *Certhidea olivacea* for a strict diet of insects.

In this brief review of the feeding habits of the fourteen species of Darwin's finches, it is remarkable indeed that among this compact group we find species adapted for feeding in ways that completely different groups of birds are adapted on the mainland of Central and South America. The needle-like bill of warblers, an adaptation for insect-eating, has evolved on the Galapagos in the bill of the *Certhidea olivacea,* while the tanager type, the honeycreeper type, the sparrow type, and so forth, are also represented by the diversity of bill types to be found among the species of this remarkable group of finches. Furthermore, a foraging life such as that of the woodpeckers, which necessitates a host of highly modified structures such as chisel-shaped bill, specialized tongue, stiff and pointed tail feathers and many, many more, is characterized in the Galapagos not by island woodpeckers but by two of Darwin's finches. The Woodpecker Finch, *Cactospiza pallida* (see Figure 17) and the Mangrove Finch, *Cactospiza heliobates,* are the "woodpeckers" of the islands, but how do they manage to forage with their long chain, nose-pliers types of bills, lack of stiff tail feathers, and so on? They do so by using tools!

Tool-using Finches

The woodpecker finch spends a great deal of its time in bushes and trees, searching for its insect prey; it does not often forage on the ground like some of its close relatives. The bird frequently hangs upside down on branches and

Figure 17. Two species of finches lead a kind of woodpecker-like way of life on the Galapagos Islands. One is the Woodpecker Finch, above; the other is the Mangrove Finch.

carefully searches the bark for crevices and cracks where its insect foods may be hiding. In the dry season, especially in the arid forests, the insects take cover during the heat of the day; this is the time when the woodpecker finches are out foraging. Now it is that we may see the famous tool-using performance of this delightful little bird. It breaks off a spine from a cactus, or else it secures a bit of twig, and pokes its tool into a crevice or crack in the bark of a tree until a disturbed insect moves. Then the bird drops its tool and seizes its prey. Robert Bowman feels that the evolution of this "woodpecker" way of life evolved in this species because of the overall dry climate so characteristic of the Galapagos Islands. As a consequence of the arid conditions, the insects upon which the bird feeds hide in crevices and cracks under tree bark during the day to avoid the heat and drying conditions. Using a tool to dislodge the hidden prey would be a clear advantage for the finch, enabling it to take advantage of a food source that would otherwise be relatively unavailable to it.

Recently Bowman and his co-workers have studied the tool-using behavior of these finches in the laboratory; they have experimented with different types of potential tools and tested the effect of hunger on the use of a tool; they observed the manner of tool use and related matters. They found that these birds may threaten one another while holding a tool in their bill, although the tool-holding bird is not always the "victor" in an encounter. It may be possible that these birds learn to use tools by watching more mature, experienced individuals do so, but this point is not clear as yet. The researchers did succeed in observing a male *Geospiza conirostris* use a twig as a tool in the laboratory; this behavior has not been seen in this species in the wild. Although the tool-using behavior of the Woodpecker Finch, *Cactospiza pallida,* has been known for many years and is well documented in the scientific literature and on film, it was only recently discovered by a German expedition to the archipelago that the Mangrove Finch, *Cactospiza heliobates,* also uses a tool. You may recall that both *Cactospiza* species have a strong probing type of bill well adapted for foraging on insects.

Feeding on Blood

The finches of Darwin continue to yield new surprises as expeditions sally forth to these intriguing islands. An expedition in 1964 discovered that one species of finch, the so-called sharp-beaked ground finch, *Geospiza difficilis*, actually attacks other birds in order to drink their blood! This amazing discovery was made on Wenman Island, one of the remote, outermost of the Galapagos group. Complete documentation of this unusual behavior was made, and it is clear that the drinking of blood is not a secondary pursuit but a primary one. Some other species of birds are known to take blood if it accidentally appears in the course of some other activity. For example, suppose a bird is feeding on engorged ticks on the side of an African mammal; the bird might then drink any blood that oozed from wounds caused by the activities of the ticks. In these cases, however, the blood was not the primary interest of the tick-eating bird as far as we know. By contrast, the finch on Wenman Island deliberately seeks out its host, works its way through its feathers, pecks at the soft skin and drinks the blood that later flows from the wound. The primary target of the finch is the Masked Booby, *Sula dactylatra*, a large sea bird (relative of pelicans and cormorants) which nests in the Galapagos Islands, commonly on Wenman Island. In fact, one of the elements that may have led to such unusual feeding behavior was the presence of the Masked Booby on Wenman throughout the year.

How could such behavior originate? Robert Bowman suggests that the association between this finch and the boobies may have begun as a standard type of relationship between a bird species and some kind of host, whether another bird, or mammal or what have you; that is, the bird landed on the host to remove parasites from the outside of the host's body (removal of "ectoparasites"). In so doing, the host was relieved of possibly harmful parasites and the reliever was provided with a source of food. The next step, and Bowman points out quite clearly that this is an assumption at this time, would be for the finch to acquire a taste for blood after it had dined on blood-engorged parasites. These are in fact readily available on the boobies of Wenman Island in the form of bird flies. Perhaps by accident, say in picking off a bird fly, the finch may have damaged the tender skin of the sea bird; blood flowed from the wound and the finch drank eagerly of the unexpected food. The next step, perhaps, was for the finch to make the puncture deliberately; that is, the finch went after the blood first rather than the parasites. But why drink blood? Of what possible advantage can this be to the finches? Bowman points out that the finches get a liquid protein meal when they drink blood; they may also derive some water from this feeding habit. Very little fresh water is available on Wenman, especially during the dry season. Thus, such a feeding method could provide the finches with critically needed water, although it should be pointed out that it is not certain that the finches do in fact receive enough water from this kind of food. Darwin's finches are a truly remarkable group of birds!

Adaptive Radiation

Here, on this group of isolated islands, hundreds of miles from the nearest continent, we find a group of closely related birds doing the job, so to speak, of a number of other groups on the distant mainlands of Central and South America. How did this "adaptive radiation" come about? Darwin pondered this problem, too, and wrote:

> Seeing this gradation and diversity of structure in one small, intimately related group of birds, one might really fancy that, from an original paucity of birds in this archipelago, one species had been taken and modified for different ends.

Suppose a small flock of birds ancestral to the Darwin's finches as we know them today were somehow to make it across the open Pacific to the remote Galapagos Islands. We have no information that would suggest such a hypothetical flock would of necessity come from the nearest part of South America. They may have come from Central America or some other place, although surely from the New World because the closest relatives of the Darwin's finches are to be found in this part of the globe. What would stimulate such a small flock to head for the Galapagos, not knowing that the islands were located out in the ocean? Was the ancestral flock a small, medium-sized or large flock? Was the "experiment" tried more than once, with various successes and failures? We have absolutely no information on which to base an intelligent answer to these and similar questions.

Perhaps our pioneering flock belonged to a migratory species, and the migrants were blown off course and barely managed to reach the haven of the Galapagos. Or, the finch ancestors may have belonged to a nomadic species. A number of living bird species are clearly nomadic in their behavior. Rosy Pastors, *Sturnus roseus*, of the Indian region, several crossbill species of the northern hemisphere coniferous forests, and some finch species of the arid parts of the Australian continent are good examples. When food supplies fail, or breeding conditions are not right, the flocks of such wanderers move around in search of food or whatever is needed for breeding purposes. Perhaps a nomadic ancestor of the Darwin's finches struck out over the ocean in search of greener pastures and "discovered" the Galapagos Islands. We don't know, of course, who the ancestor was or when the event took place, but we do know that some ancestral form did make it to the islands and gave rise to the remarkable diversity we can observe today. Whether honeycreeper type, sparrow type, tanager type or what else—we do not know—but it is possible for these small birds to fly over wide stretches of ocean and not deplete their energy reserves completely. Their powers of flight are adequate for the job.

Assume now that the ancestral birds made it to the remote island group. Perhaps what follows is their fate. Assuming that the ancestral flock was fairly specialized in its feeding behavior and associated structural features, the flock might be faced with disaster if their foods were unavailable on the islands. If they were not overly specialized, they might be able to survive if the environment of their new home could provide foods reasonably close to their normal diet. Two birds that live on the Galapagos today can serve as examples of how this might happen. Both species are flycatchers, and both species normally live in arid zones outside the Galapagos. Upon arrival in the islands, these two flycatcher species could manage to hang on because of the widespread occurrence of arid areas in the archipelago. There is no question, however, that the environment of the Galapagos would be different as compared to the mainland environments (or wherever the ancestors came from), simply because no two places on earth have exactly the same physical makeup. And it must be remembered that for living things, other forms of life are critical parts of their environment.

Thus, the ecology of the pioneers would be dramatically changed, and new conditions of life in the islands would be harshly selective. Those individuals that could survive on the islands under such severe conditions of life would be the ones that would contribute most to the next generation. Natural selection would tend to eliminate those that did not have the necessary adaptations, or that had adaptations inferior to those possessed by other members of the pioneer flock. But selection is not just eliminative, not just negative. Selection is also very positive in that the selected survivors would pass on their superior heredity to their progeny; these in turn would be subject to further selection, and gradually the descendants of the pioneers would be better adapted for life on the islands. Such a process would continue with the passage of time—selection acting on the available pool of variability—so that continued improvement in the form of adaptations of survival value would be the order of things.

Heritable variation and selection would continue until the population built up to a point where competition between the descendants began to increase. They would be competitors because selection would have molded them, so to speak, to survive under the island conditions of existence. Each individual would have passed the rigorous test of survival; each individual would have the adaptations needed for the continuance of its life. Competition would be the inevitable outcome as the population size increased to the point where similarly adapted birds would be seeking the same resources from the environment. How to avoid competition would now be the chief problem.

The vegetation and food resources of the Galapagos are not as varied as those found in a tropical rain forest, but a certain variety is available for exploitation. One way to take advantage of this resource diversity is to diversify yourself. Somehow adaptations had to evolve so that competition was eliminated or at least reduced. It is generally agreed by a large number of biologists today that such adaptations can evolve in isolation. The males and females of a sexually reproducing species will, of course, tend to breed with fellow members of their population, or with members of nearby populations. If population size were small in our pioneer flock of finch ancestors, then interbreeding within the flock would be the rule since no other populations of this species would be on the islands. But what if some segment of the pioneer population were to leave the original island of colonization and strike out for another island in the archipelago? Or, suppose the ancestral flock had somehow been split up as it first approached the islands, say by a storm, and one segment colonized one island and another segment colonized a different island.

The various islands of the Galapagos differ in their vegetation, food resources, soil types, weather and numerous other factors, so that selective pressures vary from one island to another. In this way, with small populations of ancestral birds on several of the islands, selection could begin to hasten any kind of divergence that might crop up. A heavy, seed-crushing kind of bill might be a valuable adaptation in the pioneers of one island, if such tough-coated seeds were available as foods, but on another island a better adaptation might be a slightly less powerful type of bill, one better adapted for crushing the smaller and softer kinds of seeds found in abundance on that island. In any generation there would be individual and population variation with respect to the length, shape and other attributes of the bill. Those that were best adapted to dealing with the new kinds of foods would be the survivors; they would live to reproductive age and their genes would be passed on to the next generation. Thus, divergence from the ancestral condition would begin on the several islands, and differences between the evolving island populations would begin to accrue. We can diagram this stage as follows,page 67 top.

This process could now be repeated until all of the islands of the archipelago were inhabited by the descendants of the pioneer population. If you could go back in time to the islands at this stage in their evolutionary history you might find the birds resembling one another rather closely, but perhaps subtle differences might be detected when the bill and associated feeding apparatus were examined closely. This stage can be diagrammed as follows,page 67 bottom.

Selection will act to increase the adaptive fitness of the

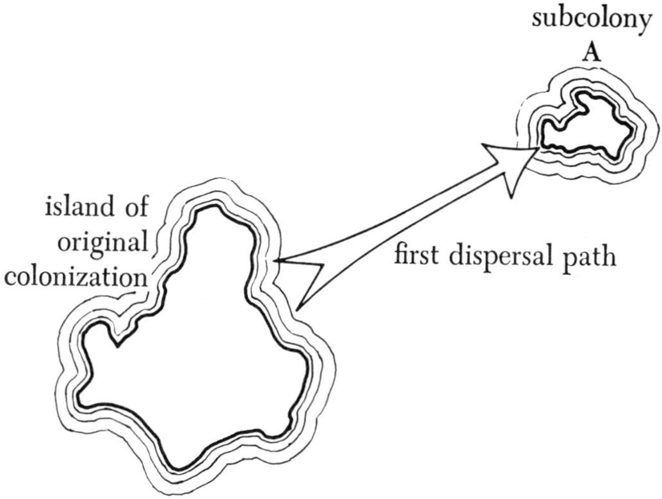
subcolony
A
island of
original
colonization
first dispersal path

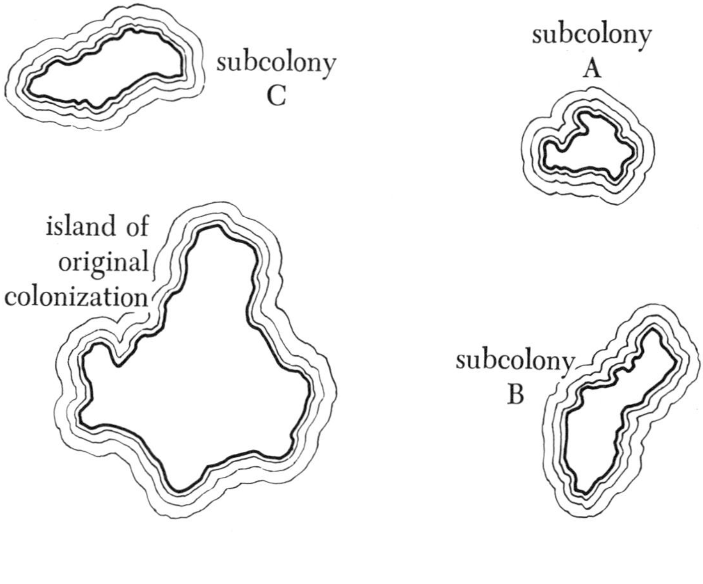
subcolony
C
subcolony
A
island of
original
colonization
subcolony
B

individuals on each island. Since the environment of each island differs, so the adaptations of the birds will differ from one part of the Galapagos to another. But have these colonists on the more remote islands reached the level of a new species? This is a valid question, one that has occupied the attention of scores of biologists since the days of Charles Darwin. When does an isolated population reach the new species level? Does it necessarily have to evolve to the new species level simply because it is isolated? Are the differences between the island populations structural, physiological, behavioral—all of these—or what?

What Is a Bird Species?

Let's digress for a bit and examine a recent definition of the species which is a reasonably satisfactory one for sexually reproducing vertebrates such as birds. This definition states: "Species are groups of actually or potentially interbreeding populations that are reproductively isolated from other such groups" (Ernst Mayr). Here is an example: The Mourning Dove, *Zenaidura macroura,* is a common breeding bird over much of the United States. If you study the breeding behavior of these birds (say on a small farm in central Georgia) you will find that male and female doves tend to mate with individuals from the same population, the local doves of that area. Perhaps your farm dove population is located some miles from the nearest dove population, but not far enough that a male from one population (or a female) couldn't fly to that nearest neighboring population and join it as a breeding member. Thus, we find that rather than being uniformly distributed over the United States, mourning doves are found in groups or clusters, populations living in habitats within which they can survive and reproduce. Interbreeding normally takes place between members of the same local dove population, but any dove of reproductive age can or is potentially capable of interbreeding with any other dove. Chances are, however, that a dove will mate with a member of its local population. There is little chance that a dove from Georgia will mate with a dove from Oregon, but it is potentially capable of such interpopulation breeding.

Now, living in many of the same habitats as the mourning doves in the eastern part of the United States is a bird called the Eastern Meadowlark, while in the western United States the very closely related Western Meadowlark lives in many habitats alongside the mourning doves. Meadowlarks have a very broad distributional range like the doves, they are subdivided into breeding populations like the doves, and a male or female eastern meadowlark has the potentiality for interbreeding with any other eastern meadowlark of the opposite sex. The same holds true for the western meadowlark. But, eastern and western meadowlarks never interbreed with doves. Meadowlarks and doves have their own distinct, exclusive kind of interbreeding. They never exchange genes with the other kinds of birds. The Eastern Meadowlark is one species, *Sturnella magna,* the Western Meadowlark is another, different species, *Sturnella neglecta,* while the Mourning Dove is a third, distinct species, *Zenaidura macroura.* The eastern and western meadowlarks are very closely related, as can be seen by their classification in the same genus (*Sturnella*), but they are distinct species. These three species are reproductively isolated; a complete breeding barrier exists between the mourning dove and the eastern meadowlark and between the mourning dove and the western meadowlark. No one has ever found a dove-meadowlark hybrid. No one has ever seen meadowlark males courting female doves or male doves courting female meadowlarks. A term often used in this context is that the Eastern Meadowlark and the Mourning Dove act as "good species," but this term is unfortunate in that it suggests the presence somewhere of "bad species." The fact that in the east the meadowlark and the dove live and breed practically side by side (or the meadowlark and dove in the west, too) in the same habitat without ever exchanging any genes leads to a technical term for designating such a relationship: scientists concerned with such problems call them *sympatric* species.

Sympatric means "same homeland," and refers to the fact that during the breeding season the species under study occur in the same habitat at the same time and do not interbreed. It does not matter really what they do outside the breeding season since no genes are being exchanged at this time. What is important in sympatry is that the species are isolated reproductively. If selection is acting on the variability of the meadowlarks so as to better adapt this species to its environment, then there will be no interference so to speak from dove genes, and the same goes for the dove population—no interference from lark genes.

Another useful and related term is *allopatric*, meaning "separate homeland." To cite a gross example, giraffes in Africa and sloths in South America are clearly allopatric, but no one would suggest that if they were sympatric they would exchange genes by interbreeding. There is no question that giraffes and sloths are distinct species. However, the situation is complicated when we are dealing with allopatric populations of very closely related forms. (See Figure 18.) Let us use as an example a species whose world

Figure 18. *Young Galapagos Great Blue Heron (above, in nest) is a geographic variant of the wide-ranging species* Ardea herodias. *By contrast the Galapagos Green Heron* (Butorides sundevalli) *(below) is considered a species distinct from its close relatives on the mainlands of North and South America.*

population consists of two groups, a large one living on one side of a mountain range and another large one living on the other side of the same mountain range. They are clearly allopatric in distribution. Are they in fact different species or geographical populations of the same species? If they were sympatric we could observe them and see if they interbred or not, but since they are allopatric we cannot have the test of reproductive isolation. This is a vexing problem, one that frustrates the student of animal classification. Names must be given to different species. In most cases this is a relatively easy matter, such as the case of the doves and the meadowlarks. The Mourning Dove and the Eastern Meadowlark (or Western Meadowlark) are clearly distinct, they do not interbreed when they are sympatric, and they pose no special problems in classification. But what do we do with the example above of the two allopatric populations on either side of the mountain range? Taxonomists (students of classification) would now have to resort to criteria other than reproduction because this test cannot be made when the populations in question are allopatric.

When faced with this kind of problem, the taxonomist turns to other sources of information: comparison of structural similarities and differences between the allopatric populations; analyses of behavior, biochemistry, etc. In short, he takes all available information at his command and then makes a decision. If the differences are great, he calls the two populations separate species; if the differences are not great in his judgment, then he calls the two populations geographic representatives of the same species. But doubt must always creep into the mind of a taxonomist under these circumstances. Where to draw the species boundary, how to evaluate his data—these are but a few of the problems he must solve. If the populations could somehow merge, come into contact, become sympatric, then the test of reproductive isolation could be taken. If they freely interbred, then the two populations did belong to the same species. If they remained isolated reproductively, then the two were in fact separate species according to the species definition given above. Although borderline cases may be frustrating and a source of irritation for the classifier, they are extremely helpful to the student of speciation because they are the examples that provide us with information as to how new species evolve.

Let's return now to our finches on the Galapagos Islands. Suppose we are at the stage of several subcolonies or island populations other than the original island of colonization. Are the populations on the outlying islands distinct enough when compared to the source island birds to call them different species? They are all allopatric and no test of reproductive isolation in sympatry is available. If we could go back in time with a practicing taxonomist as a field companion, he would be able to examine members of the various populations found on the different islands and make some kind of decision. His judgment would be based on whatever information was available to him, structural, behavioral or whatever. It would be most helpful, however, if the descendants of the outlying island pioneers were now to return to the source island from whence their ancestors came. Now they would be sympatric with the descendants of those pioneer birds that did not leave the source island but remained there and evolved with the passage of time.

Students of Darwin's finches believe that this is probably what happened in the history of the group. Allopatric populations were evolving on the outlying islands while the source island birds were evolving in turn on the original site of colonization. Now, a segment of one of the outlying populations returns to the source island. If the allopatric population had diverged enough during the time of isolation, then the segment that returned to the source island might act as a distinct species. The source island birds and the returned outlyers would be sympatric; if they did not interbreed then a new species would have evolved. If divergence had not progressed far enough in isolation then the returned outlyers would probably interbreed with the source island birds. Any differences that had cropped up in isolation might be "swamped" by the continued exchange of genes. But what prevents the outlyers from interbreeding with the source island birds, if in fact they do not exchange genes when they are sympatric? The answer is that a number of isolating mechanisms act to hinder or prevent the exchange of genes. What is an isolating mechanism?

IV

Isolating Mechanisms in Birds

An isolating mechanism is any device that prevents or hinders the interbreeding of populations that are sympatric. It is quite true that the genetic basis of isolating mechanisms may have begun in isolation, as with our finches on the outlying islands, but the true test of their efficiency or lack of it comes during sympatry. What is the nature of these devices? Let us look at some kinds of isolating mechanisms from the bird world.

Ecological Isolation

An example of habitat (or ecological) isolation is provided by two closely related species of towhees, large ground-feeding birds that are often seen scratching among the leaf litter on the ground. The two towhees, the Brown Towhee, *Pipilo fuscus,* and the Abert Towhee, *Pipilo aberti,* have overlapping breeding ranges in a small part of Arizona near Tucson. Here, the Abert Towhee inhabits the dense mesquite, willows and cottonwoods along rivers and streams. By contrast, in the same area, the Brown Towhee can be observed in much less dense vegetation, more open and free of vines, than that inhabited by its close relative. Another example: on the island of Tenerife, in the Canary Islands group off the western coast of Africa, a closely related pair of species we have referred to before in another context can also serve to illustrate habitat isolation. The birds are the Chaffinch, *Fringilla coelebs,* which is an inhabitant of lowland deciduous forests, and the Blue Chaffinch, *Fringilla teydea,* a close relative that inhabits high mountain pine forests on the same island.

The vireos of the New World, small song birds that somewhat resemble wood warblers, also exemplify habitat or ecological isolation in a number of respects: 1) some species occur in different habitats in the same region, as in the examples just cited; 2) some species may occur in the same habitat but are "stratified"; that is, one species may forage in brush or low trees or in the lower branches of tall trees, while a related species found in the same area may forage in the crowns of taller trees; 3) closely related sympatric species may show altitudinal isolation, one form occurring at the base of the mountains in the lowlands, the other high in the mountains, resembling the two Chaffinch species in this regard. It is of great interest to note that two species showing ecological isolation where they occur in sympatry may actually overlap in their ecologies where they do not occur together. Two jays of Nayarit, Mexico, occur in both mangrove and evergreen thicket habitats where they do not overlap; however, where they are sympatric, one species is restricted to the mangroves while the other is found in the evergreens.

Seasonal Isolation

If two closely related species were to breed at different seasons of the year, or were to have their peaks of breeding activity at different times in the same season, then they could be effectively isolated. Two gulls found in Europe, the Herring Gull, *Larus argentatus,* and the Lesser Black-backed Gull, *Larus fuscus,* may nest on the same island, but they tend to reach the peaks of their respective breeding seasons about one month apart. In sea birds such as the far-ranging whalebirds, or petrels, shearwaters and the like, a group of closely related species show quite different seasonal breeding. Two species may nest on the same island but at different times of the year; thus, they are effectively isolated. Males of one species cannot court females of the other species since they are not in reproductive condition and are not present on the island.

Behavioral Isolation

In many species of birds behavioral barriers to interbreeding—to the exchange of genes—are the most important of the reproductive isolating mechanisms. Two species may be weakly isolated ecologically, or they may show considerable overlap in their breeding seasons; however, no gene exchanges take place because of various behavioral incompatibilities. In birds, it is typically the male that actively searches for and courts a potential mate; he may even show his courtship displays to a female of another species. The threshold for mating is usually higher in the females; hence, although they are courted by a male of a different species, the female will not respond because the courtship displays of the foreign male are not of the right kind, are not performed in the correct sequence, or are not sufficient to arouse her to the point of mating. Although animals are capable of communicating in a variety of ways, birds tend to confine such behavior to two sensory modes: visual and aural. With the advent of the tape recorder and devices for spectrographic analysis, we now have at our disposal a large store of information on auditory communication in birds. Much of this material came from researches on how sounds could act as effective reproductive isolating mechanisms in birds.

In the Sierra Nevada of northeastern California live three species of small fly-catching birds: Hammond's Flycatcher, *Empidonax hammondii;* the Dusky Flycatcher, *Empidonax oberholseri;* and the Gray Flycatcher, *Empidonax wrightii.* The trio of small insect-eaters are what evolutionary biologists call "sibling species"; that is, they are such remarkably close look-alikes externally, at least without a detailed examination in the hand, that they are often confused in the field by human observers. Although the three species do show preferences for different habitats and could be ecologically isolated, it was found upon careful examination that there are areas where these varying ecologies bordered upon one another. Thus, the three species can be found in the same region at the same time. Experiments conducted in the field suggested that the males of the various species could not tell males of the other two species apart by structural clues. Hence, it was believed that voice might be important in species recognition, especially by the females. This line of research proved to be fruitful: all the flycatcher species show clear-cut differences in a number of vocalizations, but especially in the all-important advertising songs of the males. Thus, a female of any of the three species can easily tell the males of her species from the males of the other two even though all may be singing in the same region at about the same time.

A similar situation exists in the forest thrushes of eastern North America. The Veery, *Catharus fuscescens;* the Hermit Thrush, *Catharus guttatus;* the Olive-backed Thrush

Catharus ustulatus; and the Gray-cheeked Thrush, *Catharus minimus,* are look-alike species, but they have enough structural distinguishing features that an experienced field ornithologist can tell them apart. However, the sexes of these four species cannot be separated easily on these grounds. How do females select the proper mate when all four thrush species may be sympatric in certain parts of the United States? Experimentation and observation in the field by William C. Dilger showed that the hostile behavior of the four species was more or less similar, and this kind of behavior did not seem to provide any basis for recognition of the right mate by the females (although this cannot be ruled out completely). The vocalizations, by contrast, including the important advertising songs, again proved to be the key to effective isolation of these closely related birds. (See Figure 19.)

Visual displays can also serve to isolate sympatric species whose ecology and seasonal breeding may overlap in the same area. Two species of night herons occur sympatrically in parts of the United States. They may begin nesting at the same time, and the males of the two heron species may be found exhibiting display behavior in neighboring trees; thus, ecological and seasonal isolation can be ruled out under these conditions. At first the males are quite hostile in their response to any night heron that lands near their display perch. Soon, however, the persistent return of the females wins out and the males begin to court them. The male Black-crowned Night Heron, *Nycticorax nycticorax,* begins his courtship by doing a little "dance" in which he raises first one foot and then the other; at the same time he begins to stretch his head and neck forward. His deep-red eyes seem to bulge from their sockets. In a few seconds, the male begins to arch his neck, with his bill beginning to point down towards his toes; at the same time

Double Wing Flashing (shown by *C. fuscescens*).

Single Wing Flashing (shown by *C. ustulatus*).

Ordinary Tail Raisin of *C. guttatus*

More aggressive Tail Raising of *C. guttatus*

Two *C. fuscescens* involved in dispute utilizing Horizontal Stretch and Withdrawn postures. Insets are relatively less aggressive.

he continues his dancing steps. Soon his head and bill are below the level of his feet. At this stage the male emits a very special call—difficult to hear unless the observer is close by—one that is not heard under any other circumstance. When the "song and dance" are done, the male briefly preens his belly feathers and then returns to a normal perched posture. (See Figure 20.) At this time of the breeding cycle, the male Black-crowned Night Heron is a most beautiful species, one with lustrous, deep-salmon-colored legs and feet, shiny plumage, and blackish skin surrounding the eyes and contrasting with the deep-red irises. It is believed that the color changes enhance the effectiveness of the male's display and help him to advertise to the females that he is a male in possession of a territory and ready for mating. What about the other night heron in the next tree?

Male Yellow-crowned Night Herons, *Nycticorax violaceus*, are also brightly colored at the start of courtship, and they too are hostile to close advance by any other night heron. When the male Yellow-crowned Night Heron begins his courtship performance, however, it is almost the reverse of the Black-crown's display. The Yellow-crown male bends low on his display branch, starts to extend his bill forward, makes a little bow, and then suddenly throws both wings out to the side and points his bill upwards. At the same time he utters a peculiar whooping call not heard at any other time. In other words, the two key displays of the night heron males are opposites of one another. Charles Darwin called such displays "anti-thetical" behavior. These courtship displays, so strikingly different in

Figure 19. A number of thrushes (Catharus) *of eastern North America have similar visual displays, but these species are effectively isolated reproductively by key differences in vocalizations, especially their advertising songs.*

Figure 20. *The courtship dance of the male Black-crowned Night Heron. Males of the closely related Yellow-crowned Night Heron have a very different courtship performance.*

two very closely related forms living in the same habitats and breeding at the same time, can serve to isolate the species reproductively. Not only do the males differ in general and specific aspects of plumage and coloration, but their courtship performances are opposite in form and execution.

Suffice it to say that there are a number of devices, acting singly or more likely in concert, that can serve to keep sympatric species from breeding with one another and prevent the exchange of genes between evolving populations. Let's return now to our consideration of Darwin's finches. On our assumption that the segment of differentiated finches on the outlying islands have evolved to the point where they can now act as distinct species when they return to the source island, that is, they are reproductively isolated, then we might judge that the matter is wrapped up. Such is not the case.

In order for these two forms to act as distinct species they must be reproductively isolated and they must show ecological compatability. In other words, in order for both to survive and flourish on the source island as two distinct species, they must be able to share the resources of the island. If they do not, then the two newly evolved forms will compete directly. It is an axiom of evolutionary and ecological biology that no two species can coexist in the same habitat at the same time if they have identical needs, if they must extract the same resources from the environment. The usual outcome of such competition is that one of the species has a slight advantage in, say, foraging adaptations, and wins the competition. Or, one species solves the problem, so to speak, by leaving the situation, going elsewhere in an attempt to survive. A third solution is that natural selection may bring about such divergences in foraging adaptations that both species can survive and flourish in the same environment. One might specialize in taking hard seeds, the other in taking soft seeds. One might specialize on large seeds for its feeding needs, while the other forages for small-sized seeds. One might shift to an insect diet while the other remains a seed specialist, and so forth.

It is now believed by many biologists that the adaptive radiation of the Darwin's finches probably took place along the lines just outlined: waves of finches flew from the larger, more centrally located islands of the Galapagos archipelago out to the peripheral islands, the outliers. Then they settled there in isolation and differentiated with respect to structure, behavior and other attributes. Later, segments of these outlying populations returned to the central source islands where the test of sympatry could be met. In this way, over the years, the finches as a group evolved seed-eaters, insect-eaters, flower-peckers, and finches that acted like woodpeckers! To use Thomas Schoener's term, it was probably a matter of "ecological release" that enabled the finches to radiate. Landing on what may have been a pretty vacant group of isolated islands and not finding any real competitors such as those found on the mainland, the finches could have evolved to fit the vacant or near-vacant ways of foraging and feeding. On the mainland of Central or South America the ancestral birds were probably fairly specialized because of the presence of other, well-adapted species living ways of life not followed by the ancestors of the Darwin's finches. But on the Galapagos the ancestral flock (or flocks?) were "released," ecologically speaking, from this restriction and were free to evolve, to radiate, to occupy the various foraging modes of existence found on the remote islands. An examination of the one species of the Darwin's finches found on distant Cocos Island adds evidence to the story. On Cocos the single finch has not radiated into a variety of species filling different jobs. Why not? Populations of the original Cocos settlers had nowhere to go to isolate themselves so as to start on the path towards differentiation and subsequent speciation. Cocos Island is a single island, not a group, or archipelago. Spatial isolation seems necessary for such initial differentiation to get a start, and the needed isolation was not available to the first finch settlers on Cocos.

V

Other Islands and Continental Bird Evolution

What about birds on other islands, other archipelagos? Is the situation on the Galapagos unique? No, archipelagos in other parts of the world are just as productive of new species, new radiations as are the Galapagos. An archipelago is not even necessary, although it is more productive of adaptative radiations. All we need for speciation to take place is repeated invasion of some single, isolated oceanic island by a species of a continent. Here is but one of many instructional examples. Norfolk Island is a single, remote, isolated island about eight hundred miles east of the Australian continent. It has a very small bird fauna, but this includes three species of "white-eyes": *Zosterops albogularis; Zosterops tenuirostris;* and *Zosterops lateralis*. Where did the white-eyes come from? An examination of Australia far to the west reveals that *Zosterops lateralis* is widespread on the continent (see Figure 21), and it is reasoned that the three species evolved as follows on distant, isolated Norfolk Island. First, a segment of mainland *lateralis* managed to reach Norfolk, perhaps as a windblown group of migrants, or as a group whose navigational guidance went astray. On Norfolk, with its different environment as compared to the Australian mainland, selection would act on the immigrants and their descendants to adapt them to the new conditions. Given enough time, the pioneers would diverge from the parental stock back on the continent. Now, the accidental arrival of white-eyes from Australia is postulated to have happened once more; a second wave of birds arrives on remote Norfolk Island. If enough time has passed and enough differentiation has taken place, then the new arrivals from Australia could coexist with the original pioneers as two distinct species if they meet the tests of sympatry: reproductive isolation and ecological compatability. In view of what we find on Norfolk Island today, this test was indeed met. But yet a third wave of white-eyes made it to Norfolk; however, this most

Figure 21. The Gray-breasted White-eye (Zosterops lateralis) *is widespread in Australia but also occurs on tiny, remote Norfolk Island some 800 miles to the east.*

recent stock has not had the time to differentiate enough in isolation so as to be considered a new species. When compared with the mainland white-eyes, the most recently arrived ones on Norfolk are considered to be members of the same species, although some slight divergences have taken place when the mainland and island forms are compared.

Thus, on lonely Norfolk Island today we have three species of white-eyes, all reproductively isolated and ecologically compatible. Norfolk is large enough and diverse enough to support three species of these birds. This phenomenon is called "multiple invasion," and it is simply a smaller picture of what happens on archipelagos such as the Galapagos where waves of finches invaded the outlying islands of the group, differentiated in isolation, and then re-invaded the central source islands repeatedly, producing the adaptive radiation that we can study today. Numerous examples of such multiple invasion are now known for a host of animal groups.

Radiations on Madagascar and the Hawaiian Islands

Brief mention must be made of two additional spectacular radiations in the bird world, the vanga shrikes of Madagascar and the Hawaiian honeycreepers. The vanga shrikes are confined to the island of Madagascar, now the Malagasy Republic, a huge island located off the southeastern coast of Africa. (See Figure 22.) The vanga shrikes are said to be an endemic group, that is, they are found nowhere else in the world but on the island of Madagascar. Similarly, the Darwin's finches are endemic to the Galapagos Islands and Cocos Island. On this mountainous island off the African coast the vangas underwent an adaptive radiation similar to that of the finches of Darwin.

As with the Galapagos group, the vangas show their greatest diversity in bill length and shape, but differ in that they show striking colors and plumage patterns. In one respect, however, the vangas have not "carried" their radiation as far as the Darwin's finches. Recall that in the latter group we have both plant- and animal-eaters, but the vangas are all animal-eaters. Some species are insect specialists, while others feed on a variety of amphibians and reptiles. All of the vangas are arboreal, seeking out their foods in the treetops or from the limbs and trunks of forest trees. This, again, is in contrast to the Galapagos condition where the finches feed on the ground extensively as well as in shrubs or tall trees.

A good question at this point is: If the Cocos Island finch did not adaptively radiate, why did the vangas since they, too live on a single island and not an archipelago? The answer is that Madagascar is a very large island, one of the largest in the world; hence, evolving populations had the opportunity to become isolated from the parental stocks, a necessary prerequisite to speciation. The adaptive radiation of the vangas is not quite so dramatic as that of the finches on the Galapagos, perhaps because there is a good-sized bird fauna on the island of Madagascar. Perhaps many of the ecological opportunities available on the large island were already taken, and the evolutionary potential of the vangas was restricted by the superior competitive abilities of the island residents. Where did the ancestors of the vangas come from? Again, as with the Darwin's finches, we do not know for certain, although the vangas of Madagascar show affinities with the helmet shrikes of Africa. Perhaps the latter group were ancestral to the vangas, but nothing definite can be stated at this stage of research into this most interesting island group of birds.

The Hawaiian honeycreepers are another example of

Figure 22. *Madagascar's Vanga-shrikes* (Vangidae) *have experienced, within certain limits, adaptive radiation. Distinctive color patterns as well as differences in body size, beak shape, and diet differentiate the vangid genera.*

what can happen in the course of evolution of a group of birds that manage to arrive on and survive in the environment of an archipelago such as the Hawaiian chain. Like the vangas and unlike the Darwin's finches, the honeycreepers are brightly colored birds of crimson, orange, green, olive, and so forth. But it is the bills of these birds that are the most striking feature of their anatomy: short and straight; short and curved; long and decurved (see Figure 23); heavy and stout. One species has the tips of the mandibles crossed, somewhat resembling the crossbills of northern evergreen forests. The honeycreepers of Hawaii are classified in a family of their own, the Drepanididae, endemic to the archipelago. Students of the group subdivide the honeycreepers into those species that feed primarily on nectar and those that take foods other than nectar, such as insects and seeds. The seed-eaters tend to have heavy, seed-cracking types of bills, while the nectar-feeding species have, in general, long, curved bills with slender tubular tongues, brushy at the tips, for feeding on liquids.

Figure 23. The Hawaiian honeycreeper (Vestiaria coccinea); *total length, 6 inches (152mm).*

Where did the ancestral stock come from? What kind of birds were ancestral to the honeycreepers? Again, we do not know for certain where or from what group the marvelous Hawaiian birds descended. Perhaps a tanager-like ancestor or a warbler-like ancestor from Central America was the progenitor of the group as we know it today, but nothing is certain about the origin of the honeycreepers. What we can surmise is that the original stock, finding a relatively uninhabited archipelago, underwent an ecological release and radiated to fill the seed-eating, nectar-feeding ways of life available on the islands. Because the Hawaiian chain is a long, strung-out archipelago, the original birds were able to move from one island to another; the isolation necessary for genetic changes to come about was provided by the arrangement of the islands. It is believed that speciation went along much as it did on the Galapagos, with multiple invasion playing its part in bringing about groups of sympatric species reproductively isolated and ecologically compatible on the various islands. Unfortunately, a number of the Hawaiian honeycreepers are now extinct, and other species are quite rare, so the details of their biology are either lost forever or will be difficult to obtain in the future. This is all the more unfortunate since islands and their inhabitants are such wonderful natural laboratories of evolution and speciation.

Continental Speciation

What about continents and their animal life? Do the terms endemic, sympatric, allopatric, multiple invasion, and so forth have any meaning on a large landmass such as Africa or North America? What about isolation? Can small segments of a population become isolated and differentiate in the period of isolation, then reinvade the ancestral habitats to become either engulfed because the changes were not enough or else meet the test of sympatry and act as distinct species? The answer is yes. The principles of geographic speciation hold true for continental landmasses as well as for large single islands or remote archipelagos. Let's look at some examples of how species evolve on a large continent, taking North America as our evolutionary laboratory.

The Song Sparrow, *Passerella melodia,* has an extremely broad distribution over much of the continent. It can be met with in a suburban garden in New York City, near the edges of a marshy area in Ohio, the dry desert country of southern California, or the dark, humid forests of the Pacific Northwest. The song sparrows of the southern California arid country tend to be much paler than their darker relatives from southern Alaska. In other words, song sparrows show much geographical variation, not only in their size and coloration, but also in their songs and ecology. (See Figure 24.) In fact, when ornithologists first collected in this country they naturally took the eastern forms of this species first; these specimens were examined and described as a species new to science. But as these naturalists went westward, they began to collect song sparrows that weren't quite the same as those found on the Atlantic seaboard. For example, the song sparrows of Alaska were much larger and darker, with heavy striping on the plumage. All of these differing forms were described as new species until it was realized that these "species" of song sparrows were in reality simply geographic representatives of one broadly distributed species of bird. Such a species—one that can be subdivided into a number of geographic entities—is called a *polytypic* species; that is, one having many forms. If it weren't for the intermediate populations, the song sparrows of Alaska and of the Atlantic seaboard would certainly seem to be distinct species. But these two widely separated groups are interconnected by a series of populations stretching across the landmass of the continent.

Although there is individual variation in any population, the song sparrows of Alaska resemble one another much more closely than they resemble the song sparrows of Maryland or Tennessee. Why haven't the various sub-

Figure 24. *The Song Sparrow* (Passerella melodia) *shows great geographical variation throughout its vast North American range.*

groups of song sparrows evolved to the species level? One possible answer is that this species is only recently evolved, has spread over the continent and simply has not had enough time to differentiate to the species level in various parts of its range. Although it is clear that the various groups of song sparrows differ rather strikingly from one another, this does not mean that they have reached the species level. It is particularly clear in the West that isolation has taken place because of the greater number of races of song sparrows recognized there by students of the group.

Darwin's finches were isolated by water gaps. How about the continental song sparrows? What could act as barriers to dispersal in this case? Mountain ranges, unsuitable habitats, large bodies of water—all of these could act as barriers to the exchange of individuals from one population to another. Or, evolution of a sedentary way of life could reduce the number of individuals of reproductive age that might leave one population and join another as a breeding member. This is nicely shown in a study made of the song sparrows of the San Francisco Bay area.

In general, two types of song sparrows live in the bay area: 1) the salt marsh birds; and 2) the birds of the uplands. In many places around the bay, the two kinds of habitats are more or less isolated, leading to the evolution of well-marked population differences in the local song sparrow groups. However, where the two kinds of habitats come into contact, whether by the hand of man that disturbs the environmental relationships of the two habitats, or by natural forces, the amount of interbreeding between such neighboring populations brought into contact is increased; the differences tend to disappear.

Perhaps a song sparrow from Alaska might not be able to interbreed with a song sparrow from Georgia, but in between these outliers are interconnecting populations of song sparrows. The gaps are bridged. Perhaps if the Alaska birds were to be completely isolated for a long time, say by a severe mountain uplift to the south of their range, then they might eventually reach the species level. Such cases are instructive because they can reveal how the process of speciation works. It appears clearly that the first necessary step is for a newly evolving species to spread out, to increase its distributional range. Next, a discontinuity of some kind must develop between the main populations of the species and the outlying, peripheral segments.

Like the Darwin's finches on some distant island in the Galapagos group or the Hawaiian honeycreepers on a more remote island of that chain, the distant segment of our continental species now is subject to differing selection pressures; surely it must change, must differentiate under these conditions. The behavior, ecology, structure and other attributes of the isolated segment will differ as compared with the same characteristics of the parental population. It is not known whether or not sufficient time has passed for such changes to accrue during isolation. Only the test of sympatry and reproductive isolation will give us a definitive answer, as is also true of Darwin's finches. But if we examine a host of living species, we should be able to detect various steps in the route of speciation we have proposed, since no two species evolve at the same rates and in exactly the same ways. If this is true then we should be able to find species with little or no geographic differentiation, others with fairly pronounced changes in a number of attributes, while yet others would show dramatic changes. Finally, we should encounter species that have reinvaded the parental range with varying degrees of success. Some will interbreed freely, some casually, others not at all, thus leading to the evolution of new and distinctive species. All stages of speciation, then, can be detected on continental landmasses.

Some relatives of the song sparrow, such as the Lincoln's Sparrow, *Passerella lincolnii*, show much less local differentiation than do the populations of the Song Sparrow, even though Lincoln's Sparrow has a very broad distribution over the continent. A number of other species show as marked geographical variation as does the song sparrow; that is, they are also polytypic species. Some are so diverse that the more distant populations of the species may even be considered to be distinct species and have been so described and named. If they are allopatric, however, we do not have the test of sympatry and cannot be certain about their specific status. Some student of the group must then make a taxonomic decision. Based on all of the evidence at hand—structural, behavioral, physiological, and so on— the taxonomist judges that the two forms warrant specific recognition or not, but this is a human judgment and may not reflect the real situation. Two or more distant populations of an obviously closely related group may differ strikingly, but if they become sympatric there may be quite free interbreeding. By contrast, two or more forms that are quite similar in appearance may remain reproductively isolated when they become sympatric. Perhaps in this case there was great behavioral divergence in isolation but little or no structural differentiation. Each case must be decided on its own characteristics and not necessarily those of another group.

An interesting aspect of differentiation in structure, behavior, and such, when two closely related forms become sympatric is called *character displacement*. Many times a group of organisms will appear to have a puzzling type of variation: they will be rather similar in appearance over much of their range but quite variable in one part of that range. Close analysis may reveal that the actual situation is this: the group is in reality two aspects that resemble one another very closely where they are not found together (where they are allopatric), but they differ sharply where the ranges of the two species overlap (where they are sympatric). An example of this displacement of characteristics is to be found in two species of birds called nuthatches found in the Middle East. One species of Rock Nuthatch, *Sitta tephronota,* ranges widely to the east of Iran, while its close relative *Sitta neumayer* ranges to the west of that country. They resemble one another very closely except in Iran where their ranges broadly overlap. Sharp displacement has taken place in bill size, coloration of the head region, and so forth. Undoubtedly these differences in bill size tend to reduce competition between the two closely related species as far as their feeding ecology is concerned. Outside the region of sympatry the bills of the birds need not diverge since they are not competitors with one another. The difference in facial stripe pattern in Iran may well have to do with visual recognition signals, but little information is available for a better assessment of the significance of this difference between the two species.

The importance of differences as visual recognition signals appears to be borne out in a case of character displacement recently reported in two cuckoo species from the Philippine Islands. Here, both species have apparently similar ecologies, but there is a sharp difference in plumage that seems to be linked with visual displays. Why should two species show such differences in the area of overlap if not related to ecology? Evidence increasingly shows that hybrids produced between two closely related species are at a disadvantage when compared with either of the parental forms. Hence, selection will act to produce sharp differences in signaling characteristics so that recognition of the correct species will reduce or eliminate the production of hybrids. The phenomenon of character displacement is an important aspect of speciation biology and one that merits close attention from students of evolution.

Numerous examples are now available of how some barrier may have literally split the range of an evolving species so that the divided segments evolved on their own during the period of separation. Later, when the barrier was lifted, so to speak, the previously separated populations were free to re-invade one another's ranges. Some did so freely and interbred, but others show a most interesting pattern of limited hybridization in a very narrow zone of overlap. For example, glaciation seems to have brought about two "refugia" for birds in Europe, a pocket in the southwest and another one in the southeast. When the ice retreated, the birds in the two pockets, or refugia, could then expand their ranges and eventually meet. In the case of two crows in Europe, the Carrion Crow, *Corvus corone*, and the Hooded Crow, *Corvus cornix*, the zone of overlap and subsequent interbreeding covers a broad area, but only along a rather narrow belt of interbreeding from Scotland to Denmark, then south to the Mediterranean Sea. This belt of hybridization is a very narrow one, and outside of it the two crows are quite easily told apart in the field. Some authors consider the two crows to be distinct species, while others feel that both are members of one highly variable species. What you do with the two forms is a problem in classification, but it is a most instructive example from the standpoint of speciation. It reveals how such a situation was probably brought about by the splitting of a former broad, continuous range by the advance of the ice sheets. Similar cases are known for a number of bird species, including several from the United States.

One of the most instructive of such examples has been

studied recently in great detail by Lester Short; it deals with the flickers of North America. Flickers are woodpeckers adapted for a diet of insects taken largely from the ground, especially ants. Thus, these birds are common in open country close to trees suitable for nesting, but they are not found in thick forests with heavy ground cover. Until the recent detailed studies were made, it was considered that there were two distinct species of flickers common in much of the United States, an eastern species, *Colaptes auratus*, the Yellow-shafted Flicker, and a western species, *Colaptes cafer*, the Red-shafted Flicker. The common names come from the coloration of the underside of the wings. A third species, *Colaptes chrysoides*, the Gilded Flicker, has a much more limited range, but it is of importance in the story of flicker evolution. The eastern flicker has yellow wing linings and a black mustache stripe in the males, while the western species has salmon-pink-colored wing linings and the male of this form has a red mustache stripe.

The great artist-naturalist John James Audubon was the first to discover hybrids between the two "species" of flickers; he discovered them while collecting near Fort Union on the Missouri River in 1843. Audubon described the hybrids as a distinct species, but he agreed that they would puzzle naturalists for a long time. Actually, these two forms of flickers hybridize over a huge area, ranging from British Columbia to Alberta in Canada, then south in a great belt across the plains to Texas. It is believed that glaciation was important in separating the evolving populations, as with the two crows of Europe. The ice retreated, letting the two populations rejoin, then there was further separation by later advance of the ice, and so forth. Both forms can exist in the plains area because of the stands of trees along the water courses. In this regard man may actually be aiding the hybridization process by his own activities in this region of the country. By planting trees along the rivers and by controlling damaging prairie fires, man provides a number of potential nesting sites for the flickers.

Short's recent studies indicate that from the belt of hybridization, genes are spreading from the eastern to the western form, and vice versa. Such a process of gene-spread is called "introgressive hybridization," and a number of examples of this phenomenon are available from the bird world in addition to that of the flickers. What might be the advantage of having a "mixed bag" of genes as a result of hybridization and introgression? Some feel that because of the severity and rapid change of environmental conditions during the ice ages, it would be of advantage to be able to adapt rapidly to such changes. Introgression, it is argued, would provide a storehouse of different gene combinations that might give the carrier of these gene combinations an advantage under conditions of stress. Probably it is too soon to make any final statements about the evolutionary advantages or disadvantages of such mechanisms of gene exchange, but it is a most interesting phenomenon that has opened up new lines of study.

In summary, the evidence suggests that—at least for vertebrate groups—some degree of isolation, some degree of apartness, is needed for different gene combinations to be built up in populations that are geographically separated. Since the populations are separated in space, then their environments will be different. Thus, the individuals in each population will be subject to different kinds of stresses from both the physical forces of their environment and the other living things seeking to survive in the same surroundings. With the passage of time slight changes in structure and how they work will accrue. Eventually, differences in ecology and behavior might be noted as well as more obvious differences in structure. If these evolving, isolated populations are successful, then they will begin to expand their ranges as population size increases. Finally, the two populations may meet one another; separateness now becomes living in the same area together. If the structural, ecological and behavioral differences that evolved in isolation are great enough, then the two populations may act as distinct species when their ranges do come to overlap. If so, then they do not interbreed and they utilize the resources of their common environment so that little or no competition occurs. But, the story is not finished. Much remains to be done in this ever-fascinating realm of how new species appear on the earth and evolve. Birds have contributed much to our understanding of these problems, and they offer rich rewards for future research along these lines.

Part Two

*

MAN AND BIRDS
Behavior

VI

Territory

History of the Concept

The concept of territory in animal life has had a long history, and the study of this kind of behavior in birds has contributed greatly to our understanding of the phenomenon. An ancient Greek who mused about territoriality in birds was the philosopher Aristotle. This Greek naturalist noted that a pair of eagles seemed to require a large space for themselves; in addition, he observed that the large birds of prey excluded others of their kind from their domain, more or less forcing them to occupy neighboring areas. The word territory as used in a modern biological sense was introduced by the English writer Oliver Goldsmith in 1774. Goldsmith used the term with reference to the behavior of small songbirds which appeared to "mark out a territory to themselves, which they will permit none of their own species to remain in; they guard their dominions with the most watchful resentment; and we seldom find two male tenants in the same hedge together." Thus far, territorial behavior seems to be concerned with male birds setting aside a plot of ground for themselves, which they defend against other birds of their own kind.

The essentials of modern theories of territorial behavior, however, at least for birds, were first stated clearly by the German ornithologist Bernard Altum in 1868. In essence, Altum argued as follows: 1) individual pairs settled at fixed distances from one another; 2) the male drove other males of the same species from his territory; 3) the function of male song was to threaten rival males of his own species and to attract females; and 4) the main function of the territory was to provide food for the adults and their young. Although Altum's ideas on this most interesting and important facet of bird behavior were well known and respected in Germany, they were virtually unknown outside of his native land. The whole subject of territoriality was brought to the attention of the world's biologists by an English ornithologist, H. Eliot Howard. Completely independent of Altum, Howard formulated his ideas on this kind of behavior in a series of books published between 1907 and 1920. In the latter year he published his now classic book *Territory in Bird Life*.

It is one of those unusual coincidences that two men in different countries think alike on some important scientific idea, yet neither one is aware of the other man's work or thoughts. Howard's ideas are essentially a restatement of those of Bernard Altum. In summary, Howard said: 1) male birds isolate themselves in the spring (at the start of the breeding season); 2) the male birds restrict their activities to a well-defined area (the territory); 3) the male birds are intolerant of trespass on their domain, especially trespass by males of their own species; 4) the male bird proclaims possession of his territory by means of loud song, which functions to repel males of his own kind but also serves to attract females of his species to the territory. Howard concluded that the function of the territory was to supply food. Howard's little classic stimulated a rash of research on territorial behavior in birds and other animals, and the results of this scientific effort revealed general agreement with Howard's findings (and those of Altum) except in one major respect: the functions of territorial behavior. Today we have a most impressive list of possible functions for this fascinating field of animal behavior.

Some researchers felt that this behavior functioned to limit the size of a population, because only a certain number of males would be able to secure a territory in which reproduction would take place. Others believed that the function of this behavior was mainly as an anti-predator device. The male and his mate, according to this line of reasoning, would become quite intimate with the vegetation on their territory and could quickly duck for known cover in the event that a hawk or other possible enemy were to appear suddenly on their territory. Some felt that Howard's points about male song and its functions were most important in an analysis of such behavior; that is, they concluded that the territory provided a song center for the males as well as an attraction area for the females. Reduction in rate of transmission of diseases via the spreading out of the individual territories was another suggestion for the overall function of territorial behavior. And so it went; all of these proposed functions for this behavior indicated that Altum and Howard had hit upon an idea of great biological significance. It was a concept of basic importance, and one worthy of very close attention and renewed research on a variety of birds and other animals.

The English Robin

Since birds were the first objects of study with respect to territorial behavior, let's examine in some detail the various features of such activities in several unrelated kinds of birds. First, a species that is not familiar to most Americans, but would seem to be by virtue of its common name —the English Robin, *Erithacus rubecula*. This species is extremely popular in England and in Europe. It has a bright orange-red breast, and it is a readily seen, highly active species that invites further observation. (See Figure 25.) Early immigrants to America, seeing the brick-red

Figure 25. The English Robin (Erithacus rubecula) *has been much studied with respect to its territorial behavior. Note the leg rings or bands for individual identification.*

breast of the American Robin, *Turdus migratorius*, dubbed it "robin," even though the two species are not that closely related. The American Robin is a thrush, a relative of such fine songbirds as the Veery, the Hermit Thrush, and the Wood Thrush, while the English Robin is more closely related to species not as familiar to Americans as the forest thrushes. Pair formation in the English Robin takes place between late December and March in the latitude of England. This period, especially in late December and January, is noted for the renewed and vigorous singing of the male robins. (See Figure 26.) After courtship and mating, the female robin builds a nest during March and lays her eggs; eventually the young birds fledge. During July and August the adults are rather shy in their habits as they are undergoing a molt of their plumage. The pairs separate at this time, and the males adopt a territory in which they sing and chase other robins from their domain. This behavior, however, tends to weaken as the fall of the year passes. But then in late winter, as noted above, there is renewed singing. Some female robins also adopt their own territories in the autumn and winter which they defend against others of their own kind, but some females migrate to winter quarters to the south.

Figure 26. Vigorous singing by the male is characteristic of the period of pair formation in the English Robin.

While on their winter territories some of the female residents sing as do the males, while other females remain silent. After pairing with a male late in the winter, the female is rarely heard to sing. In territorial defense, by contrast, the female of a mated pair is just as vigorous in her aggressive activities as is her mate. Both sexes are similarly colored and cannot be told apart in the field. Thus, to obtain the information given above, it was necessary for the researchers to live-trap the robins and affix small, colored bands or rings around their legs. This technique allows for the gathering of a tremendous amount of information on the behavior of known individuals. The standard government band provides information as well, but it does not individualize each ringed bird. Using this method of coloring individuals, the researchers determined that males and females are equally aggressive in fighting and defense of the territory. Further information thus provided was that song was more frequently given by the males and that separate autumn and winter territories were held by both sexes.

When a rival robin trespasses on a territory, the male resident challenges the newcomer with a song, which may or may not intimidate the intruder. If it does not, then the resident flies to a perch close to its rival and displays its bright orange-red breast directly at the trespasser. (See Figure 27.) This is a most striking display, and female English Robins display in precisely the same way as their mates. It is of interest to note that when an intruder trespasses in search of food, it may be threatened and attacked but will return, perhaps more boldly this time, to renew its search. You will recall that Howard said that one of the prime functions of this behavior was to provide a food supply for the robins and their young. Detailed studies revealed, however, that robins frequently trespassed for food, even though they vigorously defended their own

Figure 27. *Threat displays of the English Robin. Note how the posturing emphasizes the bright orange-red breast which is directed at the intruder (on the right).*

territories. This finding suggested that a food function was not the complete story in this species. One researcher suggested that perhaps the territory was more important in providing a place where pair formation could take place without interference, and the pair bond could be maintained throughout the breeding season on a commonly used, familiar plot of ground. Thus Howard's concept of the function of territorial behavior was being expanded in the light of later researches.

The American Robin

A look at the territorial behavior of the American Robin will also show that Howard's conclusions about the main function of this behavior may not hold true for a large number of bird species. American Robins are highly migratory, and it is the males that arrive first on the breeding grounds. Shortly after their arrival, the males tend to localize their activities, and they become very aggressive towards other male robins. Since the females have not arrived as yet, the fighting here is definitely concerned with the territory, and not a mate. Threats and fights are very common at this stage of the annual cycle, and fighting between several males will attract others to the fray at times resulting in "gang" fights. A bit later the females arrive; the female robins seek out the males on their territories and mate with them. Like their English cousins, the female American robins are most vigorous in defense of their mate's territory. In fact, the female may be even more aggressive than her mate. There is extensive trespass on neighboring territories to the point where the owners of a territory that has a rich source of food may very well be overwhelmed by the extent of the intrusions. In this case defense of the territory is not feasible. Much food is gathered by the robins on their territories, but common food grounds are not unusual. Thus, the main function of territoriality in the American Robin does not seem to be to provide a food supply for the adults and their broods. One researcher, Howard Young, concluded that some functions of this behavior in the American Robin were to provide a place in which mating and rearing of the brood could take place as well as providing some food. Let's turn now to a consideration of a quite different species whose territorial behavior is strikingly different in at least one respect; that is, the food aspect of such behavior.

The Green Heron

The Green Heron, *Butorides virescens,* is a common member of its family found over a broad area in the United States south through Mexico and Central America to the Panama Canal Zone. I have had the opportunity to study this bird in a number of places in the United States as well as in southern Mexico. I began my studies in the salt marshes of Jamaica Bay, Long Island, New York. Here I found that this species is a migrant, arriving on the islands in the bay about the middle of April to start a new breeding season. This small heron migrates singly or in small to medium-sized flocks that contain both sexes. Upon arrival on the nesting grounds, the males at once move into the trees and reeds and start to advertise their presence in this habitat by flying about, singing throughout the day, and chasing and threatening each other. (See Figure 28.) By contrast, the females spend the daylight hours out on the vast marshes that surround the nesting areas. Here they forage for hours, leaving only to spend the night in the roost.

After the initial flurry of activity the males settle on in-

Figure 28. Early territorial activities of male Green Heron: singing (above) and low-intensity threat (below).

dividual territories which they vigorously defend against intrusion by other male Green Herons. A great deal of their time during the daylight hours is spent giving their advertising song, which is a low-pitched, rather harsh-sounding "skow." Most often the male is perched high in the top of a tree while he gives his advertising "skow," and he is most conspicuous against the sky. But sometimes the male will utter his song while he walks about on the ground, searching for suitable nesting materials. In the Green Heron it is the male who begins construction of the nest; some males finish the nest, while others let the female complete the task after the pair has formed.

After a few days of this preliminary threatening, chasing, advertising and occasional fighting, the males have carved out for themselves a rather large area whose boundaries are well known to their neighbors. At about this stage in the cycle, the females are attracted to the territories of the males for the first time. The first response of a male resident to the presence of a female on his territory is an aggressive one. He first threatens the female and then launches into flight and drives her away. Females, however, seem to behave similarly in a number of unrelated species under these circumstances. They are persistent; they return again and again to the male's territory, even though they are repeatedly repulsed. Soon, however, their persistence pays off. The males shift from aggressive to courtship behavior and the rest of the cycle soon follows: mating, completion of the nest, egg-laying, and rearing of the brood. Rearing of the brood may terminate reproduction for that year, but some pairs rear a second brood before flying south towards their wintering grounds.

In general, my findings on this bird agree with those in studies of the English and American robins. The territory of the male Green Heron provides a place in which he can advertise for a mate, court an interested female, and maintain a pair-bond with her for the remainder of the breeding season. Similarly, the female Green Heron assists the male in defense of the territory but not nearly as vigorously as either of the robin species females. However, it is in the food function of the territory that the Green Heron differs so strikingly when compared to the robins. No food whatever is gathered on the territory by the Green Herons, so this function of territoriality as claimed by Eliot Howard does not apply to this species in any manner. (See Figure 29.) Where do Green Herons obtain their necessary foods? They forage and gather their prey items—mostly small minnows—out on the vast saltwater marshes some distance from the breeding territories. It is of interest to note that out on these marshes, foraging Green Herons will threaten another individual that approaches too close. Since I could not tell individuals apart out on the marshes, I do not know whether or not an individual returned to and defended the same feeding area day after day. It is quite clear, however, that this species does not gather food on its nesting territory.

Figure 29. *It was believed that one of the classic functions of bird territoriality was that it provided food for adults and their young. However, research on the Green Heron (shown above in a nest relief ceremony) indicates that no food at all is gathered by adults on the nesting territory. Rather, a separate feeding territory is defended.*

With findings such as these on a large number of bird species, it was generally agreed that some elements of the ideas of Altum and Howard on territorial behavior were quite sound, while the food function was clearly in doubt in a number of cases. After World War II, an important meeting was held in England, a gathering of students of birds interested in problems of territoriality. An attempt was made at this congress of enthusiasts to come up with some general statements about the nature of this all-important aspect of bird behavior. The first order of business was: how could territory be defined? There was general agreement that a definition of territory that had been given some years before the congress by G. K. Noble was suitable. Noble stated that territory could be defined as *any defended area.* Note that this definition does not imply that there are any rigid boundaries to the defended area, nor does it suggest that the defended area must be a place where reproduction takes place. The defended area or territory could be a breeding area, a roosting site, a wintering territory, separate foraging area, and so forth. However, it was agreed that in general the defended area was some kind of localized, topographically located place, but defense of a mate or some other object that could move outside the defended area would be included in the definition.

Types of Breeding Territories

Breeding territories are subdivided into four major categories, following a scheme proposed by the American ornithologist Margaret Morse Nice some years before the British meetings. She recognized a Type A territory, which was a large breeding area within which courtship and nesting and a large part of the foraging and food-gathering took place. This is the now-classic territory of Altum and Howard. The English Robin and its American cousin, discussed in some detail previously, are good examples of birds having a Type A defended area. Secondly, the Type B species maintain a large breeding area in which courtship, mating and nesting take place, but little or no food-gathering is observed. The Green Heron is a good example of such a species. Colonial-nesting birds such as gulls and terns, as well as many other kinds of birds, exemplify species that hold Type C territories. Here, a small area immediately around the nest is defended by the resident. Courtship and mating may or may not take place on the territory itself, but the defended area clearly provides a nesting site and a place to rear the young. In such territory holders all food is gathered away from the nesting site. In Type D the birds have a defended area which is used strictly as a mating station; no nesting or rearing of the young or food-gathering is involved in the functions of such a territory. Included in this category are the famous "lek" birds, and their behavior on the mating station is so interesting that a digression is in order. What is a "lek" bird?

Lek Birds

Lek birds are those species in which the males move to a traditional, communal display ground early in the breeding season, threaten and fight one another there, and try to secure a display station so that they can mate with the females that are later attracted to the display arena by the vigorous activities of the males. The display area, or site, is typically called a "lek," a word coming from the Swedish language. The word in Swedish means game, or play, with sexual overtones. Lek species are found in a number of diverse bird groups: gamebirds such as grouse; some species of hummingbirds; some birds of paradise; a few shorebirds such as the European ruff, and so forth. Typically the males of a lek species are easily told from the females. Males have strikingly different coloration, plumage, size and highly specialized display behavior. The males of such species have nothing to do with nest building, incubation of the eggs or care of the young. All parental care devolves on the females; they are typically colored in a protective manner, an adaptation of obvious survival value to the female and her brood. As soon as a female has mated with a male on the lek, she leaves the display grounds and goes about the tasks of rearing the next generation. The males remain on the lek waiting for more females to be attracted so that they can be courted and their eggs fertilized. Let's look now at some of the activities of the males on this specialized type of mating station; we will use several outstanding examples recently studied in some detail.

The Black Grouse, *Lyrurus tetrix*, is a game bird (relative of pheasants, quail and similar birds) found commonly in a number of places in central and southern Europe. (See Figure 30.) Males are larger than females and distinctively colored in a deep bluish-black; they have a striking, lyre-shaped tail. The communal, traditional display grounds or leks of this species are typically found in open, grassy country, the more open the terrain the better. The lek is used throughout the year, but more often in the spring and summer during the peak of reproductive activities. Males arrive at the lek before dawn and begin to display quite vigorously. They make great use of both visual and auditory behavior. The calls of the males resound through the area about the display ground as they suck air into their esophagus and then expel it to produce the loud vocalizations. Vigorous challenges are met by counter displays or fights as the males struggle for a place on the lek. Espe-

Figure 30. Black Grouse (Lyrurus tetrix) *males gather on traditional communal display grounds called "leks" where they challenge and fight each other for the favored centrally located display sites.*

cially in favor are the sites near the center of the display arena. It has been found that on some leks only a handful of the cocks (as few as four) effect 85 percent of the matings with the females attracted by the displaying males.

Turning now to a different part of the world, we find a most interesting species of lek bird in the forests of the tropical island of Trinidad. Here, the Black and White Manakin, *Manacus manacus*, holds forth on its communal display grounds. (See Figure 31.) Again, as with the Black Grouse and many other lek species, the males and females are quite differently marked. The males are a rather showy black and white, while the females are a dull olive-green in coloration. Both sexes feed on fruits, which may be an extremely important fact in the evolution of lek behavior in manakins. The males spend most of their time on individual courts or display sites on the overall lek. Such courts are used year after year by the male manakins. Females visit the courts, mate with the displaying males, and then

Figure 31. *Manakins are lek birds of the New World tropics. The showy males have nothing to do with parental care, a typical trait of lek species. Shown is the male Yellow-thighed Manakin.*

leave the lek. All parental care is left to the females. In order for full display to be shown by a male, he must have at least two slender saplings present on his court. His displays are performed while he is perched on or sliding down the saplings. In order for a communal lek to become established, there must be a large number of such slender trees spaced but a few meters apart over the display area. Perhaps twenty-five or more males will be seen displaying on the same lek. The Black and White Manakin males have a number of specialized feathers in their plumage. These are used to make unusual snapping noises and whirring sounds during bouts of displaying. Like the Black Grouse, there is much jockeying for a display court near the center of the lek among the manakins. If a central court resident is temporarily absent, a more peripherally located male will enter the vacant court and display therein until the resident returns and terminates the intruder's behavior.

The leading student of manakin behavior, D. W. Snow of England, feels that the evolution of such behavior in these birds rests on two main points: 1) that they have a fruit-eating diet; and 2) that the male is freed from all parental care. A fruit-eating diet seems to be essential because the time needed for searching for foods is much less if these dietary items are fruits rather than insects and similar prey. Hence, the males can spend a great deal of time on their courts, leaving but briefly to seize and devour some nearby fruits. Freeing of the male from all parental duties necessitates having a mate that can build a nest by herself, incubate the eggs alone, and care for the developing young without any assistance. Similarly, a fruit-eating diet makes these duties possible for the female to handle, especially the feeding of the developing brood.

There are many other fascinating species of lek birds—such as the bizarre birds of paradise of New Guinea and neighboring islands, the booming prairie chickens of the North American grasslands, the amazingly variable male ruffs of Europe—but we must return to our consideration of the general aspects of territorial behavior. At the meeting of students of this kind of behavior, there was no general agreement about its major function or functions except for the widespread feeling that it was probably different for every species! For some species, such as the English Robin and the Green Heron, the territory does in fact provide a place where courtship, mating and nesting take place; the territory of the English Robin additionally functions to provide some food for the adults and the young. By contrast, the territory of the lek birds seems to function solely as a mating station. Each species must be studied in turn to find out what functions are subserved by its kind of territoriality. Generalizations based on the study of a few species—such as the generalizations made by Altum and Howard—may not hold for many other kinds of birds. What about territorial behavior not connected with reproductive activities?

Other Kinds of Territories

As noted some birds such as the English Robin may maintain separate winter territories. Mates of just a few weeks ago may now be challenging trespass on each other's territories in the late autumn and winter. Mockingbirds, *Mimus polyglottos,* in the United States may adopt such winter territories; at this time of the year an observer can easily see vigorous challenges, threats and even fights almost daily. One possible function of winter territories is that they may provide critical food needs during periods of hardship. Some evidence for this view comes from a number of bird studies. For example, the Red-headed Woodpecker, *Melanerpes erythrocephalus,* a bird of the deciduous forests of eastern North America, maintains a distinct winter territory. Much of the bird's daily activity is devoted to foraging for acorns. The nut-gatherers store their acorns in caches under the bark of trees. Such supplies for hard times are vigorously defended, not only against other Red-headed Woodpeckers, but also against such potential acorn-eaters as jays and titmice.

A most interesting kind of territory, held throughout the year, is that defended by the European Nutcracker, *Nucifraga caryocatactes.* We had occasion to refer to this species earlier when we were discussing ways of caching food supplies as a hedge against hard times. This bird, a member of the crow family, was studied in detail in Sweden by O. Swanberg. These birds pair for life and set up a territory in the conifer forests, a domain on which they remain permanently. The conifer forest is typically located near a hazelnut woods, a most important feature of the nutcracker's life. During the early autumn and until the snows begin to fall, the pair of nutcrackers spend virtually all of the daylight hours gathering hazelnuts in the deciduous woods. They bring the nuts back to their territory in the conifer forest and bury them in holes that they excavate in the ground. A small pile of nuts is placed in the cavity and then is covered over. Later the hidden stores are further covered by snow. The nutcracker pair nest early, while the snows are still on the ground, but the birds feed themselves and their young on the stored nuts. Swanberg was able to observe the pair digging through the snow in order to retrieve their stored supplies. They were able to do so even through a foot or more of snow cover. Remarkably, the nutcrackers rarely made a mistake and dug down to a cache that they had already utilized. It is believed that a detailed knowledge of their territorial terrain, brought about by permanently residing in one area, enables the nutcrackers to recover their stored supplies. Each boulder, fallen tree, and other feature is known to the mated pair, a knowledge that is essential to the survival of their young. In the United States, it is suspected that the closely related Clark's Nutcracker, *Nucifraga columbiana,* may engage in similar caching of supplies for later recovery, but little is known about this aspect of their behavior.

Group Territories

So far we have been describing examples of territories defended by a male alone (lek birds) or at best a mated pair (English Robin and other species already noted). However, some species of birds show *group* defense of a common territory. A fine example comes from Australia. There, a crow-like bird called the Australian Magpie, *Gymnorhina tibicen,* exhibits social territoriality. Robert Carrick, the leading researcher of the species, was able to determine that from two to ten magpies lived in a group, the smaller number always being made up of one male and one female. But the composition of the larger groups would range to three adult males, three adult females, and up to seven immatures in the flock. Territory size also varied, from five to twenty acres being defended by different groups in the study area; the smaller-sized territories were found in regions where pressure from nearby magpie groups was greater. Attack and defense in this bird species is always done by a group, as a team effort and not a solitary one. Carrick was able to record that there were several kinds of magpie groups: 1) permanent groups that live on their territory all year round and do all of their life activities such as gathering food, nesting and roosting, on the group territory; 2) marginal groups that had territories in poorer habitats with little in the way of food supplies; 3) mobile groups that met vigorous opposition from the holders of the permanent territories; and 4) open and flocking groups that do not attempt to breed. It is of interest to note that the only successful breeding is done by the holders of the permanent territories; in fact, Carrick believes that only about one-quarter of the adult population actually engages in breeding. So, territorial behavior in this species seems clearly to act as a controller of population size. Group territoriality is also shown by the Smooth-billed Ani, *Crotophaga ani,* a member of the cuckoo family. In this species groups of fifteen to twenty-five birds defend a common area that contains a communal nest, one in which several female members of the group lay their eggs. Young birds sometimes remain with the group the following year, rather than disperse and found new groups.

Interspecific Territoriality

In our descriptions of territorial behavior to this point, we see that defense of an area is normally confined to intraspecific activities; that is, posturing and related hostile or aggressive behavior is directed towards a member of the same species as the displayer. In recent years, however, a number of research projects have focused attention on another aspect of territorial behavior: interspecific, or between-species interactions. Here the territorial behavior involves two or more species, sometimes quite closely related ones. Work on several blackbird species in North America has proved to be quite rewarding in this respect.

The Red-winged Blackbird, *Agelaius phoeniceus*, the Yellow-headed Blackbird, *Xanthocephalus xanthocephalus*, and the Tri-colored Blackbird, *Agelaius tricolor*, are blackbirds involved in these studies. (See Figure 32.) The Red-winged Blackbird has a very broad distribution over much of North America, with recent nesting records as far north as Alaska. This is the blackbird with the red "shoulder patches" so familiar to many schoolchildren throughout the United States. The females of this species, however, are a dullish, heavily streaked brown. The Yellow-headed Blackbird is broadly sympatric with the Redwing over much of the western United States. The Tri-colored Blackbird has a very restricted range along the west coast from southern Oregon and northern California southward through the lowlands of California to northernmost Baja California in Mexico. The Red-wings and Yellow-heads are highly migratory species, while the Tricolors tend to be nomads, although the exact nature of their nomadic behavior is not known in detail.

Although the Red-winged Blackbird is often characterized as a "marsh-nesting" species, a careful examination of

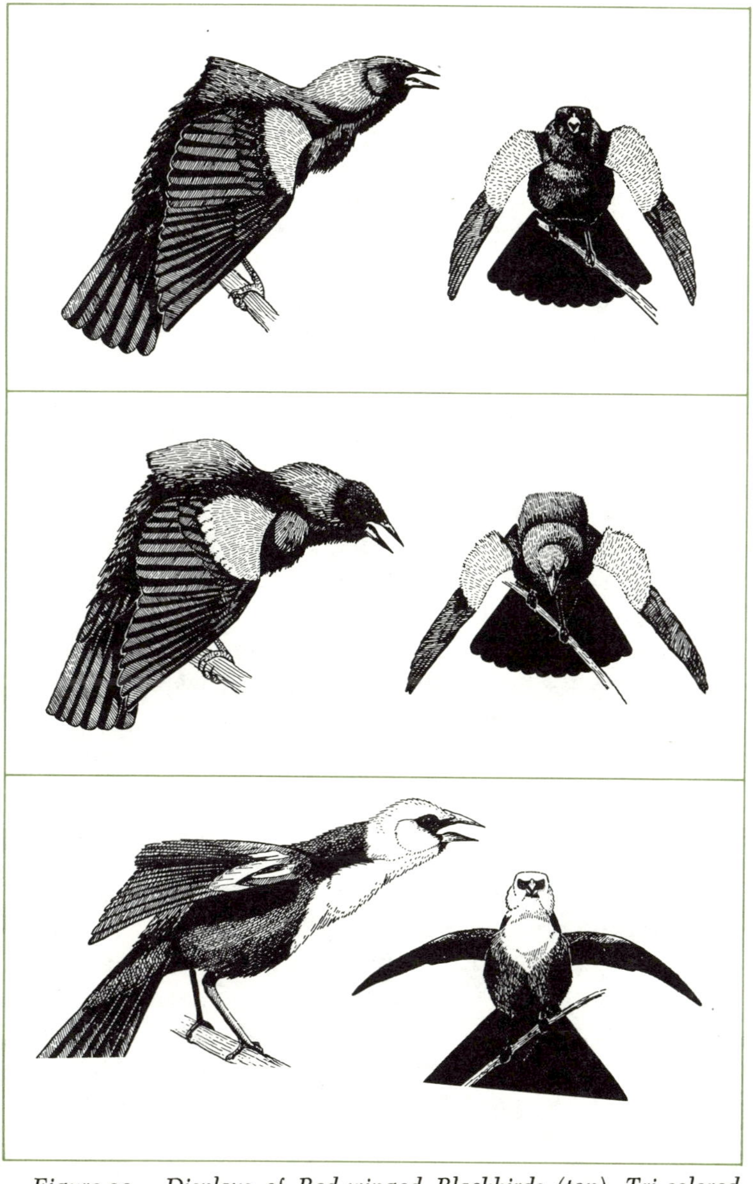

Figure 32. *Displays of Red-winged Blackbirds (top), Tri-colored Blackbirds (middle), and Yellow-headed Blackbirds (below). These three closely related species are often involved in interspecific territorial interactions.*

its nesting habitats reveals that it is much more broadly adapted than that designation would suggest. For example, Red-wings will nest over standing water in marshes (which seems to be their preferred nesting site), but they will also nest in willows near the water's edge, in trees and bushes near the water, or even several miles from the nearest aquatic situation. By contrast, the Yellow-headed Blackbirds are strictly confined as breeding birds to marshes and sloughs with a great deal of emergent vegetation such as cattails, tules, rushes, and the like. The Yellow-heads build their nests over the water and have not been observed to nest in any situation away from standing water. This clear-cut difference between these two species is important in the interspecific territoriality shown by them.

In many parts of their ranges where the Red-wings and Yellow-heads nest sympatrically, it is the male Red-wing that arrives first on the breeding grounds, while the females of this species arrive much later. This would seem to give the male Red-wings a clear-cut advantage in securing a suitable nesting territory. However, this does not seem to be the case. It has been found out that the male Red-wings take much more time in establishing their territories than do the male Yellow-heads. As soon as they arrive on the breeding grounds from migration, the male Yellow-heads establish and begin to maintain their territories, while the male Red-wings, although having arrived several weeks before, are quite slow to begin. This is clearly shown in parts of California where the male Red-wings are non-migratory. Here, however, the females do vacate the breeding grounds and migrate southward in the autumn. In such places in California, the male Red-wings may actually settle on a territory in January, but the females do not arrive from the south until March. Thus, a seeming advantage of early arrival on the territory is not an advantage when the behavior of the males is examined critically. Another factor is the clear-cut dominance of the Yellow-head males over the male Red-wings in those marshes and sloughs that have an abundance of emergent vegetation. One study in eastern Washington revealed that all of the preferred sites near the center of the marsh used as a study area were occupied by Red-wing males in late March. When the Yellow-head males arrived, however, they evicted the Red-wings from the choice territories in the center. By late April no male Red-wings were in residence in the preferred center of the marsh. After losing the "battle of the center" to the Yellow-heads, the male Red-wings, or at least some of them, settled on less desirable sites at the edges of the marsh. Some moved to another habitat for nesting.

The territorial behavior of the Tri-colors is very different when compared with that of the Red-wings. While the male Red-wing will be on his territory for weeks or even months before nesting begins, the male Tri-colors do not settle on a territory until the very day that the first breeding begins. These nomads settle and breed only in those areas where a suitable food supply is available. Colonies of fifty to one hundred thousand birds have been observed. The nesting of such hordes of Tri-colors is highly synchronized. In one study area, for example, all nests in one huge colony were finished within one week of arrival of the birds on the nesting grounds. The territory of the male Tri-color is also much smaller than that of the male Red-wing. Furthermore, the Tri-colors do much more feeding away from the territory; at times they may be found foraging three or four miles from the breeding sites.

Many of the visual displays and the vocalizations made by these three blackbird species are strikingly similar (see Figure 32), but there are some notable differences that seem to be correlated with the nesting habitat and kind of territory found in the three species. For example, flight displays are much more common and easy to observe in the male Red-wings and Yellow-heads than in the Tricolors. No doubt this is related to the fact that these two species have larger and more open territories where visual activities would be more easily seen. The red shoulder patches of the male Red-wing and the bright-yellow head of the male Yellow-headed birds would seem to be much more conspicuous in aerial displays over the marsh vegetation. The small territory of the male Tri-color, by contrast, would not allow for aerial displays because of the unceasing trespass this would bring about. Hence, we find that there has been an evolutionary "shift" towards the greater use of guttural vocalizations in this colonial species. As more and more birds are studied in greater detail, it is becoming clear that much of the behavior of a species is related to the presence or absence of closely related species. These relatives may compete for the same or similar territories. Additionally, a great deal of territorial behavior is clearly related to the nature of the habitat in which the birds must display: over water, in a forest, and so forth.

Recall that the definition of territory as "any defended area" did not specify that there were any rigid boundaries to such a defended place. Some species do in fact have rather well-defined borders to their territories, while other species are rather flexible in this respect. At times the pressure from newcomers may cause a previously vigorous defender to yield so that the borders of his domain shrink. This will enable the newcomer to carve out a territory for himself. In my own work on the Green Heron an interesting change in behavior of the earliest male arrivals on the breeding grounds is related to this point. Late-arriving males were able to secure a territory for themselves as follows: the first males to arrive on the study area did so on or about April 15th (recall that the study area was in New York City). They at once entered the nesting area and began to advertise their presence, threaten and chase one another, and so on. Soon these earliest arrivals are settled on individual territories. The borders of these areas are well-defined and the resident male vigorously challenges all trespassers. I was able to make a detailed map of the different territorial boundaries by paying special attention to the places where actual contact fights took place.

In addition, I noted that when a resident male would leave his territory temporarily, neighboring males would take advantage of this situation by trespassing and attempting to gather nesting materials from the ground, steal a twig from the resident's nest, and so forth. The interesting point I want to make in this regard, however, is that when the intruders crossed what I took to be the boundary of their respective territories they began to flip their tails up and down in a "nervous" or tense manner. Additionally, such intruders erected and depressed their dark crests rapidly; this action I was able later to determine indicates a tense Green Heron. Even though the boundaries were not marked in any way, the intruders knew exactly where they were at all times.

Soon after the first wave of males in residence were well established, the females began to visit them on their defended plots of ground. At first the females fly from one territory to the next, mainly because they are chased by the resident males. After this initial greeting of hostility, however, the males begin to show very low-level courtship behavior. At this stage in the breeding cycle, the resident males will tolerate trespass on the very edges of their territories, although they will not allow penetration to the center of their domain. In a day or two the males are showing more sexual displays (see Figure 33), and their territory size shrinks a bit more. They begin to pay more

Figure 33. *Key sexual display of the male Green Heron on his breeding territory is the "stretch display," shown from the side (left) and from the front (right) as seen by the female.*

attention to the visiting females than they do to the periphery of their territories. Soon the defended area is a matter of the nest tree itself or even just the space immediately around the nest. The female is permitted to land in the crown of the nest tree; but if she walks down a limb towards the nest, she is repulsed and takes her place in the crown of the tree once more. At this stage the male is showing vigorous sexual or courtship displays; he spends very little time in threatening and chasing his neighbors.

Now the female is allowed to enter the nest and stand side by side with the male in the structure he has been working on for several weeks. This is the moment of pair formation in the Green Heron. Now the mated pair jointly defend the territory of the male, but defense is limited to the area immediately around the nest. Meanwhile, new males have been arriving from the south. These newcomers settle on initially large territories. The late arrivals seem able to recognize that the early arrivals are no threat to them since they rarely challenge the early arrivals. Perhaps it is because these early arrivals are now pursuing purely parental duties and are not advertising, threatening, chasing or fighting. Such behavior continues until the end of June. The original large territories of the first males to arrive in April may now contain half a dozen pairs or so of Green Herons, all getting along with one another very nicely. By such a shift in motivation from almost pure hostility to almost pure sexuality, combined with a staggered migration, the population of Green Herons in the breeding area can be much larger than would occur if all of the first arriving males were to maintain their vigilance and exclude all members of their species from their initially large territories.

We have been using birds as our prime examples of territorial animals, but the whole subject of territorial behavior is fraught with significance for other kinds of vertebrates. A number of studies have revealed the occurrence of similar behavior in a number of invertebrate species as well. Man, too, is included in the ranks of territorial animals. All of this research indicates that the subject is of great biological importance, but there is much more to be learned about this fascinating area of behavior. This statement holds true for birds also. Even though the phenomenon of territorial behavior was first studied in birds, much is still to be learned of this aspect of their biology, even for a large number of common "garden" species.

VII

Bird Communication

For centuries naturalists have known that a host of animal species have rather specialized structures, movements, sounds and other devices which seem to be involved somehow in communication between individuals of either the same species or different species. The growling of a dog upon seeing a strange dog on his territory, the hissing of a snake as a potential enemy draws near, the clucking of a hen to bring her chicks to newly discovered food, and so forth. There was much speculation about such devices, and more often than not the behavior of the animals being observed was interpreted in human terms: the mother bird "teaches" her young to watch out for hawks; this or that individual is "afraid" of a certain situation. Such examples, and many others, are familiar to us all.

The Releaser Concept

Actually, little progress was made in the field of animal communication until the year 1935, when an Austrian zoologist named Konrad Lorenz published a very important scientific paper in a German bird journal. Later this contribution was translated into English, and the various ideas of Lorenz began to circulate among interested students of animal communication. One of Lorenz's important concepts was that of the "releaser." What is a releaser and how does this idea help us to understand how one animal can communicate with another?

Lorenz noted, as had many naturalists before his time, that a number of devices in different animal groups seemed to be involved with communication of one kind or another. Furthermore, he noted that some of these devices seemed to trigger very special kinds of behavior in the reacting individual. Lorenz called these special devices releasers, and he defined a releaser as any structure, movement, color, sound—that is, any device that would produce a key signal to which another individual would respond in a highly predictable manner. Usually, the other individual was a member of the same species as the signaler. It might be a mate, one of its offspring, an intruder on a territory, a competitor for food, and so forth. In short, these special devices served to *release* special kinds of behavior in the receiving animal, which is why Lorenz called them releasers. Let's look now at a number of different kinds of releasers or signaling devices to see how they can help us to understand animal communication.

Visual Communication

There are, of course, many ways and numerous devices for communicating information, but we'll examine them from the standpoint of the sense organs involved in receiving the signals. Thus, we can review visual releasers, auditory releasers, chemical releasers and touch, or tactile, releasers.

Visual signals may be concerned with color, form, movement or any combination of these elements. The English Robin, which we discussed in connection with its territorial behavior, provides a nice example of how color is involved in signal systems in this conspicuous species.

Recall that when a resident male on his territory spied an intruder about to trespass on his defended area, he first sang in an attempt to intimidate his opponent. However, if this did not prove successful, then the resident would fly over towards the intruder, perch near it and then begin to posture. This display always included orientation of the bright-orange-red breast directly toward the opponent. One student of the behavior of English Robins, David Lack, decided to experiment with stuffed birds placed on the territories of resident individuals. When the stuffed mount of an adult English Robin was placed securely on a territory, say it was fastened to a branch, it was at once threatened, postured at, and then rather viciously attacked. These attacks were so violent in nature that the poor stuffed bird was ripped asunder. The wings and part of the breast were still intact in one trial, but the mount was now headless! But the attacks of the resident continued until the entire mount was wholly destroyed. Could it be that the attacks were prolonged because at least a part of the bedraggled mount still had some traces of the orange-red breast feathers left? To test this idea, another stuffed adult was secured, but this time the orange-red breast feathers were disguised by staining them a brownish color, so that the mount now resembled, in coloration at least, a juvenile English Robin. David Lack hid from view to watch the reactions of the resident bird.

This time the mount was largely ignored! Such a response, or lack of it, suggested that it was in fact the orange-red breast feathers that acted as a signal, releasing attack behavior on the part of the resident. Further experiments were tried. Just a tuft of adult breast feathers was wired to a branch on an occupied territory, and the result was most interesting: the tuft of feathers was threatened, attacked, and ripped asunder. If a stuffed juvenile and a tuft of adult breast feathers were placed side by side on a branch, the tuft was attacked and destroyed but the whole mount of the juvenile was ignored. In one experiment, with a particularly aggressive female, a stuffed adult was placed on a stick, which was rammed into the ground on the female's territory. The mount was duly threatened and attacked for a period of about forty minutes, with sustained vigor. The experimenter then removed the offending mount and the stick, but the female was seen to return to the very spot where the mount had been placed and attack the air! The orange-red breast of the adult English Robin would seem clearly to qualify as a signal functioning as a Lorenzian releaser. (See Figure 34.)

The manner in which a bird postures can provide a great deal of information to others of the same species. In other words, postures can act as visual signals. When Herring Gulls, *Larus argentatus,* are on their island territories, in a few moments an observer can easily see a fine example of the subtle information that can be conveyed by slight changes in postures. When a resident gull spies encroachment on its territory it moves at once towards its opponent. Then it stops and next adopts a very characteristic pose: its neck feathers are erected so that the neck of the posturing bird has a very thick look to it. It holds its bill at a slight downward angle, and the bend of the bird's wing is quite obvious as it holds its wings slightly away from the sides of its body. The intruder may adopt the same posture for a moment. But let's assume that the resident makes a very slight forward movement. If the intruder is at all intimidated, it will now depress its neck feathers, hide the bend

Figure 34. *David Lack clearly demonstrated that the orange-red breast of adult English Robins functioned in visual communication when he offered whole stuffed juvenile mounts (which lacked the brightly colored breast), heads, wings, and single breast feathers from adults to territorial residents. Only the mounts with orange-red on them were attacked.*

of its wing by holding its wings tightly to the sides of its body, and finally it will raise its bill to the horizontal. If the resident continues its approach, we may now see the intruder make its neck very thin by depressing its neck feathers to an extreme degree; also, this gull will hold its bill above the horizontal. These posturings indicate that the gull is in readiness for a rapid retreat. Careful observations over a long period have indicated that these slight differences in posturing can reveal information about the motivation of the posturing bird. For example, the thickened neck indicates a very aggressive gull, one that is about to launch directly into an attack; hence, it has been dubbed "the aggressive upright threat display" by students of gull behavior. The intermediate posture, with thinner-looking neck and bill held horizontally, indicates a tense or "anxious" gull, one that seems to be undecided about what to do next. This stance has been called the "intimidated upright." Finally, the gull with the extremely thin-looking neck, bill held upwards and no bend of the wing showing is signaling its intent to the territory resident that it is about to withdraw from the encounter. This latter posture is called the "anxiety upright." If an observer spends some time in a crowded gullery, he can predict—with a good measure of reliability—the outcome of such encounters because he has learned how to "read" gull language. (See Figure 35.) Many bird species show a similar graded series of threat displays and posturings ranging gradually from almost pure attack behavior to almost pure

Figure 35. Hostile behavior of Herring Gulls (Larus argentatus). *The gull on the left is very aggressive and ready to attack its opponent; the gull on the right is intimidated and is about to flee from its opponent.*

escape intention. Some species, however, show "all-or-nothing" types of hostile displays. Here the displayer suddenly shows a very intensive aggressive posture with no graded intermediates beforehand.

Movements as well as postures can also act as signals to release a response of approach, withdrawal or some other response, depending upon the nature of the signal and the signaler. For example, orientation of the body may indicate that the signaler is about to attack or flee, or may reveal a conflict in the displayer's motivation. Many birds will orient frontally towards an opponent, and at the same time they will reveal some structure that emphasizes the frontal movement. The showing of an associated structure such as a flash pattern on the bend of the wings or a differently colored crest or breast patch, helps to make the orientation movement much more conspicuous. In all likelihood, the movements and the associated colors and/or structures evolved together to make the display more effective as a signal. Sometimes the effect of the orientation is heightened simply by a fluffing of the body feathers, a movement that suddenly gives the displayer an apparent increase in size. Such "inflationary" displays often intimidate an opponent. A fine example of this kind of display is shown by the Wood Thrush, *Hylocichla mustelina*. High levels of hostility are indicated in this species by an extreme fluffing of the body feathers (see Figure 36), while low levels of hostility are shown by an extreme flattening of the body plumage, giving the bird a very thin look.

In my own researches on the Green Heron, I was able to determine that this species shows a number of postures that involve color, form and movement in various combinations. Careful study over the years gave me clues to the motivation of the displaying birds. What the bird did with its crest or tail, whether or not it raised the modified

Figure 36. Front and lateral views of high-intensity spread postures of the Wood Thrush (Hylocichla mustelina). *Note the apparent increase in size brought about by this fluffing display.*

feathers on its back, and so on, were indicators of what it was about to do next. (See Figure 37.) Male and female Green Herons are real look-alikes outside the breeding season. For example, if you were to take a boat to the mangrove islands of the Florida Keys in the winter time, you would be able to find Green Herons skulking around the prop roots of the red mangroves in shallow water. The birds at this time of the year are dull-colored, not at all striking in coloration. However, when the migratory flocks head northwards, some of the flock members are seen to show striking changes in the coloration of the bill, legs and feet, and the bare skin in front of and around the eyes. The more brightly colored individuals are the males, although the females show more modest color changes in the same areas of their bodies. Additionally, the plumage of the male is especially lustrous at this time. By the time the migratory flock arrives on the nesting grounds, it is rather easy to tell the males from the females.

One function of the brighter color changes in the males may be to facilitate recognition of the sexes, because in other respects—size, general appearance—they are similar. Furthermore, the color changes in the male may signal to the female that she is observing a male in reproductive

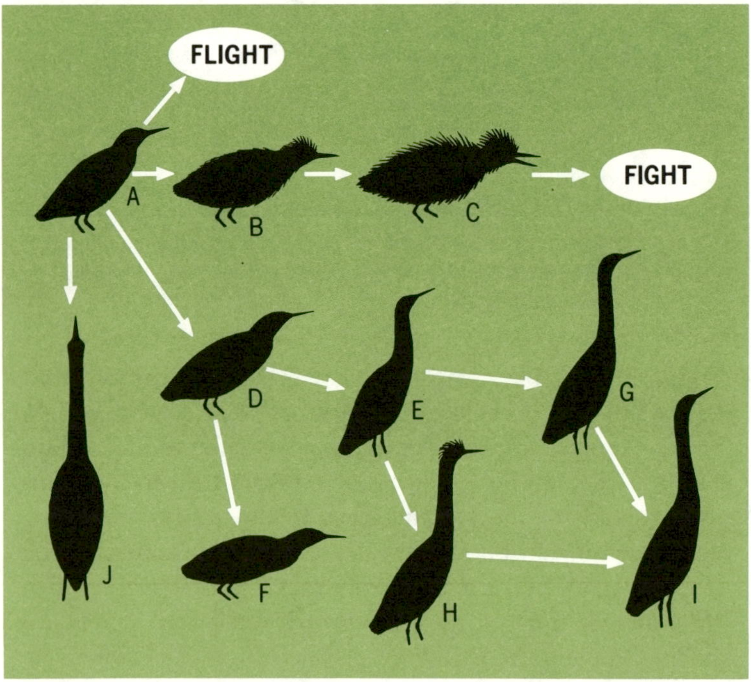

Figure 37. *Response of a Green Heron to an intruder may take the path A-B-C-Fight as shown at the top in the silhouettes above, or it may be one of a number of "anxiety" postures as shown at bottom (D-J). The latter postures are typical of responses to potential predators but not to other Green Herons.*

condition. At times Green Herons in sub-adult plumage may be found in a nesting colony; studies of some other species of birds have shown that sub-adults or adults breeding for the first time may not have as great a success in rearing a brood as more experienced adults. Such color changes may also enhance the many displays of the territorial male, making him much more conspicuous.

There is no doubt, however, about the meaning of the signal when a male Green Heron threatens and challenges an intruder of his territory. His first response to the intrusion is to adopt a low crouch with frontal orientation of his body towards the trespasser. At the same time he erects his dark crest slightly (see Figure 38), starts to flip his

Figure 38. Low-intensity threat of a male Green Heron. Note the crest.

tail and to raise the modified plumes growing from his back. If the intruder is not intimidated by this display, which is a relatively low-intensity aggressive posture, then the male shows a much more hostile form of this posture. All of the features noted above are exaggerated: extreme crest erection; back feathers fully erected; very strong tail flipping in the vertical plane; eyes seeming to bulge from their sockets. Furthermore, the male suddenly opens his mouth, showing the deep-red lining of the interior. Finally, the male gives a very harsh, rasping call in combination with the visual elements of the hostile display. (See Figure 39.)

Usually the trespasser withdraws upon seeing the resident display in this very aggressive manner. But if the interloper does not retreat, even in the face of such a vigorous hostile posture, then the resident launches into flight and either drives the intruder away or engages in a vicious fight with it. Fighting is rather rare among Green Herons, however; numerous recent studies in animal be-

Figure 39. *High-intensity threat of a male Green Heron. The open mouth indicates that this primarily visual display is accompanied by a harsh, rasping call at high intensities.*

havior have shown that a great deal of hostility does not involve actual physical contact but is more a matter of threat, of challenging and bluffing. It is easy to see why this kind of aggressiveness should have evolved rather than more damaging kinds. Actual contact fights might prove fatal or at least very damaging in the sense of a broken leg or wing. A bird with such wounds would be in grave danger; it is in no condition to cope with its predators, and its normal means of escape—flying or running or swimming—may be lost to it. Studies of the hunting behavior of some birds of prey such as falcons have shown that these predators tend to select the "oddball" in a flock for attack. That is, the individual of the prey species that doesn't fly exactly like the others in the flock, the sick or the maimed bird. Something peculiar about the locomotory behavior of this individual brings it to the attention of the predator, often with fatal results.

Although the change from low-intensity to high-intensity threat postures in the male Green Heron are rather clear-cut, the postures of alarm and anxiety are much more subtle in their signal content. A Green Heron may indicate various degrees of alarm largely through the extent to which it raises and lowers its crest and tail, and the degree to which it extends its neck. (See Figure 37, page 142.) The crest of this species is a good indicator of ambivalence: it tells the observer that the posturing bird is in conflict about what to do next, whether to flee from the source of stimulation, remain and attack, stand motionless, or whatever. Green Herons do one thing that clearly indicates they are terrified, however; this occurs when a large bird of prey such as a Marsh Hawk, *Circus cyaneus*, suddenly sweeps over the nesting area and catches the unwary Green Heron by surprise. Instantly the heron freezes in place by extending its head and neck upwards so as to appear very thin and remains absolutely motionless. This posture serves to make the heron blend in with the overall vertically aligned vegetation of reeds, cattails and the like in its nesting environment. This effect is enhanced by the brown and white vertical stripes on the heron's neck. This posture is very characteristic of bitterns, cousins of the Green Heron, so I have called it the "bittern stance" of the Green Heron. (See Figure 40.)

In the previous example of the Herring Gull's graded series of hostile postures, it was pointed out that an opponent could tell whether the posturing bird was about to attack, to retreat, or to stand fast. If the displaying bird is very carefully observed, it is possible to tell if it is in conflict. For example, if a gull in an aggressive posture walks

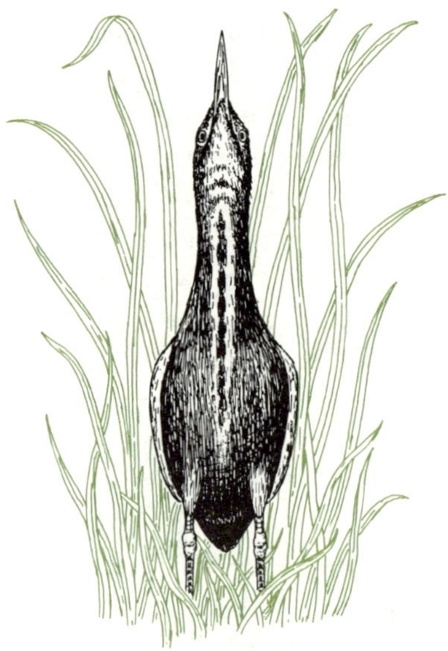

Figure 40. *"Bittern stance" of the Green Heron, given in response to the sudden appearance of an aerial predator such as a Marsh Hawk.*

slowly and rigidly towards its opponent, it is also leaving the center of its own territory. It may be about to encroach on the territory of its neighbor, the very bird it is threatening. At times such as these, we may see some subtle changes in the angles of the wings and bill, or the thickness of the neck, that tell us the aggressive gull is having another thought about the whole matter. Surely these subtle changes are noted by the other gulls; sometimes the intruder, previously showing a low-level posture in response to the challenge of the aggressor, perhaps even intimidated, may now shift to a more aggressive display and actually force the displaying bird back towards the center of its own territory.

One of the most interesting moments in a crowded gull colony comes when two males are face-to-face in a real stand-off at the borders of their respective defended areas. They may feint at one another with a quick thrust of their bill, but they can be seen to withdraw their head quickly as the opponent retaliates in kind. Suddenly, one or both of the antagonists will jab viciously at the ground, ripping up the turf. Some of the movements in this activity resemble the actions of nest-building, but the gulls are not at all interested in parental duties at this time. Students of gull biology have called this kind of behavior "grass-pulling." Other species of birds under very similar circumstances will show feeding movements, as if out of the blue, and yet others will just as suddenly appear to assume a sleeping posture. But if these displays are carefully observed, we can see that the "feeding" bird doesn't pick up any food, the "sleeping" bird is wide awake and keeps an eye on its opponent at all times, while the "grass-pulling" gull does not build a nest on the edge of its territory. These behavior patterns seem completely irrelevant to the extremely hostile situation in which they appear, so they have been called "displacement activities." It seems that in a situation of extreme tension and conflict, the birds thus involved do something that more or less resembles other kinds of behavior normally shown at different times, or in different places under other conditions. Little is known about the actual nervous processes that may bring about such displacement activities, but they are fascinating to contemplate.

So far we have seen examples of visual displays performed while the animal was more or less stationary. A threatening bird may move towards an opponent, or it may raise its wings or crest, but it tends to remain in or near one place. A number of visual signals, however, are brought about in birds by displays that are given in full flight. Herons again provide some interesting examples. Soon after the resident male Green Heron allows the female to perch on the edge of his territory without challenge, the male will begin to show the first signs of courtship, or sexual activity. From a crouched posture, while standing in or near his partly completed nest, the male will suddenly launch into flight. If the observer has seen normal flight in this species then he has noted that in a few moments or so after takeoff the heron retracts its head and neck and tucks its legs and feet neatly in line to the rear of its body. Its flight is smooth and fairly rapid; at a distance, the flight of a Green Heron tends to resemble that of a Common Crow, *Corvus brachyrhynchos.*

On the territory, however, the flight of the male as he launches from the nest platform is strikingly different. The male raises his head slightly, crest fully erect as if in a hostile display, while his legs and feet dangle beneath his body. More surprising is his manner of flight: the male literally lurches through the air like some giant moth. (See Figure 41.) He flies around the periphery of his territory, in the flight posture as described, and then he returns to his nesting tree with a jolt. His landing may actually be a crashing into the branches, and he may lose his balance on colliding with the tree and fall a foot or two. From an observation hide, I was able to get very close to such displaying males, and my only conclusion was that the male seemed to be ecstatic as it lurched through the air and came crashing back to the nest tree. At this stage of the breeding cycle, the male is brilliantly colored and has very lustrous plumage; his bright coral-orange-colored legs and feet stand out clearly as they dangle beneath him. The female watches the displaying male very intently from her perch high in the crown of a tree at the edge of the male's territory; sometimes a female will fly behind the male during a bout of such displaying.

A related species, the Snowy Egret, *Egretta thula,* has a remarkable aerial performance. I believe that this display is one of the most beautiful sights in the world of heron behavior. The male of this species is pure white with highly modified plumes growing from its back, the famous

Figure 41. Flap flight display of a male Green Heron, exhibited during the late stages of his courtship.

aigrettes of the plume-hunting days. The black legs of the male glisten, while its toes are a contrasting, shiny orange-red. Like the Green Heron, the male Snowy Egret also shows color changes in the bare skin around the eyes. In the Snowy Egret, this area becomes a bright scarlet and can change color with the shifting motivations of the male. While displaying to the female, the male Snowy Egret will raise its head up, erect its fine back plumes to the point where they resemble an ethereal bowl, then pump its head up and down vigorously and give an accompanying gurgling call. After a few such stationary performances, the male will now launch into flight and do the same thing in the air as it flies about on its territory. (See Figure 42.)

Figure 42. *Spectacular aerial courtship display of the male Snowy Egret* (Egretta thula) *shown in silhouettes that read from the lower left to the upper right as the bird launches into flight.*

This is a most amazing and beautiful performance, especially since the male keeps his head and bill pointing straight upwards during the aerial display!

Many species of birds have aerial performances as part of their visual display repertoire. Birds of the grasslands or similar open areas, because they cannot sing from favored song perches for lack of trees or bushes in their habitat, may do so from a rock or other protuberance if it is available. More often the male will advertise his presence on a territory by suddenly flying high into the sky, almost out of sight of the observer, and then give his song at the top of his climb. In some species the male will then follow this aerial advertisement with a quick dive back to the protection of the grasses. In other open-country species, however, the male may follow his advertising song with a series of spectacular gyrations, all bringing attention to himself. In most cases the female is in cover nearby, watching the displaying male with the greatest of interest. Such aerial displays may function to advertise the presence of the male on his territory, they may be hostile in nature, or they may be part of some courtship ritual of the male, an accessory to his stationary behavior. Again, as with the three closely related species of North American blackbirds, we find a close relationship between the forms of a bird's displays and its habitat type.

Darwin's principle of "anti-thetic" behavior is of importance here in a discussion of visual signaling systems. Recall the example of the Black-crowned and Yellow-crowned Night Herons mentioned earlier in the discussion of birds and evolution. These two species are very closely related; they do not interbreed with one another, a necessity for sympatric species. There is no exchange of genetic material between these two species of night herons because the key displays of the males are *opposite* in nature. When a female of one species sees a male of her own kind showing the important courtship postures characteristic for her species, she will respond to his displays and mating will take place a bit later. But if a male of another species, even a closely related one, is posturing in a nearby tree, as so often happens with these two kinds of night herons, then the female will not respond to the behavior of the "foreign" male because it does not perform in the correct manner. Selection will act to bring about a wide divergence in the courtship and related activities of such closely related forms so as to insure that matings take place only between members of the same species. Thus, the herons provide additional examples of Darwin's idea of anti-thetic behavior, "doing the opposite," as well as sharp differences in courtship between closely related forms nesting in the same or neighboring trees.

Recall that the Green Heron adopted a low crouch when an intruder appeared on the edge of its territory. If a low-intensity threat display did not suffice to intimidate the rival, then the male showed a much more hostile posture and the intruder either fled the scene or else a fight ensued. Now, when the female Green Heron first lands on the territory of the male, she will be threatened and repulsed from the defended area. The male uses the same displays directed towards the female as he does when threatening a rival male. But how does the male signal to the female that he is ready for courtship rather than a fight? He does so in several ways: 1) by showing less hostility towards the female; 2) by showing the first signs of sexual behavior via the unusual aerial display described above; and 3) by showing a display which is the *opposite* of the major threat displays. I have called this display the "stretch" display because the male stretches upward, raising his head and bill straight up; then he does a little side-to-side swaying kind of dance. Note here that the male's bill, the chief weapon in a fight, is directed *away* from the female. Furthermore, the male does not give any harsh, rasping calls as he does when most aggressively threatening an intruder. By contrast, the male utters a soft, gentle gurgling or cooing sound as he sways from side to side during the performance of the stretch display. This display clearly has a different meaning; it is a different kind of signal to the female: come a bit closer to the nest site; I will not attack you; I am ready for courtship. This is anthropomorphic, of course; but the meaning of the signal to the female Green Heron is quite clear: the male's hostility is receding and his sexual drive is rising. The other heron species that I have studied in detail all show similar anti-thetic behavior. Postures at the courtship stage are just the opposite of those but a few days before; the bristling, challenging fighter of yesterday is the cooing, gurgling courter of today. If the male failed to show such opposite types of signals, he would probably never get a mate. He must be aggressive at the start of the breeding season to secure a territory that is vital for reproduction, but later his hostility must be reduced so that he can mate with a female and continue the remainder of the nesting cycle.

A fine example of anti-thetic behavior is shown by the "hooded" gulls. Such species have in both sexes a dark, brownish-black color to the head region during the breeding season. For example, the Laughing Gulls, *Larus atricilla,* of the Florida area where I live do not have the hood coloration in the late fall and the early part of the winter, but instead have a head coloration that is more or less whitish. But in February you can see that some of the gulls are starting to show dark feathers among the paler ones. By March and April all of the adults have a fine hood, one that contrasts nicely with the snowy white and silvery gray of their main body plumage.

If you study the behavior of the Laughing Gulls or that of their very close relative the Black-headed Gull, *Larus ridibundus,* of Europe, you will soon see that the males of these two species are extremely aggressive on their territories. Also, when the female is challenged when she lands on the male's domain, she will be almost as aggressive in her responses as her future mate. Such a situation would not lead to the continued existence of either gull species unless the hostility of the "pair" could be reduced or subdued, permitting courtship to take over. This is actually accomplished by a rather neat little ceremony called "head-flagging," or "facing-away." When a Laughing Gull or a Black-headed Gull is threatening a rival, it does so by aiming its dark reddish-black bill directly at its opponent. The dark hood of the aggressive gull makes a fine facial mask, emphasizing the frontally oriented display. In the anti-thetic behavior, by contrast, the two gulls stand upright, look at each other, and then quickly turn their heads and bills *away from each other.* This movement reveals the white feathers on the back of the head and not the brown hood of the frontal, hostile posture. Furthermore, the weapon of the gull, the bill, is directed away from the other individual. (See Figure 43.)

This head-flagging, or facing-away behavior, functions as an appeasement posture since it appeases or tends to reduce the hostility of the displaying birds. Early in the formation of the pair, the two birds are under great strain for a number of reasons: they may or may not know each other from a previous nesting season; the male has been vigorously fighting for a territory; his aggressive drive may be very high, and he is apt to strike out at any other gull of his species; either bird may be breeding for the first time, and so forth. In order for pair formation to occur, the hostility of both the male and the female must be appeased so that further interactions leading to nest-building, egg-laying and care of the young can take place.

I shall defer the evolution of visual communication dis-

Figure 43. Gulls such as these hooded Black-headed Gulls (Larus ridibundus) *appease the hostility of their mates by a mutual "facing away" display that effectively averts their bills, one of the chief weapons used in gull combat.*

plays until we can have a look at sound communication in birds. At that time I shall discuss how both visual and auditory modes of signalling may have evolved. At that time we shall also examine how one change in a bird's biology—the place where it builds its nest—may have profound effects on its behavior—visual, auditory, and otherwise.

Sound Communication

Most people like to observe birds and to listen to them because birds are often colorful, they tend to be quite active—even spectacular—in some of their behavior, and they fly and make themselves conspicuous. Most bird species can be easily identified in the field with a suitable guide, and often such identification is based on some kind of sound made by the bird. It is this latter attribute that we need to examine now in some detail since we are creatures of sights and sounds, and we tend to emphasize these aspects of communication when we study other kinds of animals. Among birds, of course, sound communication is of utmost importance since chemical and touch (tactile) modes of communication are virtually non-existent or else of little importance in most bird species.

Bird sounds may be true vocalizations; that is, the sounds (songs or call notes) are produced in specialized sound-making organs, or else they may be mechanically produced. Air moving over modified feathers, tapping of the bill on hard ground or on the trunk of a tree are examples of such sounds. As far as the function of bird sounds is concerned, they may be produced in connection with foraging for food; to indicate a state of hunger; to proclaim the finding of an abundant food source; to warn others not to get too close to a food supply, and so forth. Additionally, bird sounds may warn of the approach of danger; they may be used for various reproductive purposes; they may be a proclamation of territory ownership; they may be courtship or mating calls; or they may help to keep a family group or a flock together, say on migration.

All of us have heard the food-begging calls of young birds in the nest, and perhaps we have seen the fledglings leave the same nest. At this time they may be heard to give a similar call or perhaps a different one so as to bring the parent to their location for feeding purposes. Thus, this call of the young has a dual function: 1) it serves to indicate that the young are hungry; and 2) it acts as a locator signal informing the parent where to bring the food. If the young birds were fully exposed, which would make the job of the parent easier, then the young would be vulnerable to predation. However, if they are partially or completely hidden from view, the parents can reach them quickly by means of such calls. Sometimes it is the parent that calls and attracts the young to a food source it has found. When a hen with chicks finds some food, it clucks, and the chicks can easily locate their parent by this simple sound, even though the hen may be out of visual range. Some adult birds have a special call that they give when they return to the nest with food; the function of such a call may be to stimulate the young birds to gape so that the parent can feed them.

A number of birds give harsh warning calls when another individual draws too close to a source of newly found food. Once I watched a Common Egret, *Casmerodius albus,* capture a fish that was too large for it to consume by the normal heron method of gulping it down head first. The egret would drop the fish, peck at it, hold it scissor-fashion in its bill and so on, but to no avail. Whenever another heron or egret drew too close, the owner of the fish would show a low-intensity threat posture and give a guttural "frawnk" kind of call at the same time. This was usually sufficient to make the curious intruder retreat. It is easy to hear such threat calls around a food source if you have a bird feeding station in your backyard. Some individuals can clear a feeder as they approach by uttering a harsh call just before they come in for a landing. Nearby birds scatter and let the tyrant have his fill. The Great White Heron, *Ardea herodias,* of the Florida Keys, can clear a turtle grass flat of all smaller species of herons and egrets simply by uttering its harsh threat call as it sweeps over the shallows. After the dominant Great White Heron has settled in a foraging zone of its choice, the smaller species subordinate to it may return and start to forage rather close to their larger relative. (See Figure 44.)

Just the opposite kind of behavior is shown by at least one species of gull, the Herring Gull, in that this bird has evolved a special food-finding call that is only uttered by a gull that has found a superabundant supply of food. Tape recordings made of this call by Hubert and Mable Frings and their associates were played back on a coastal strip of land to test the attracting power of the special call. Shortly afterwards, a number of Herring Gulls were drawn into the area and began searching for the food source. It

Figure 44. *Larger herons on a Florida turtle grass flat are dominant to the smaller species present. Great Blue (right) and Great White Herons can clear a foraging area simply by giving a harsh call as they approach for a landing.*

was noted by these researchers that a solitary gull was silent if it had discovered only a small supply of food. W. H. Thorpe of England, a leading student of bird sounds, suggests that such specialized food-finding calls are uttered only when a bird discovers a temporarily superabundant source of food. I made an observation in Florida that confirms this statement. I saw a teeming throng of fiddler crabs (*Uca*) on a mud flat on the edge of Tampa Bay one day in December. The milling hordes of crabs had been discovered by a small group of Ring-billed Gulls, *Larus delawarensis*. The gulls gorged themselves until the crabs entered the bordering red mangroves. Gulls will not enter such cover, so the crabs were a temporarily superabundant source of food in Thorpe's sense. I was unable to tape-record the sounds made by the frenzied gulls as they fed, but to my ear it seemed to be a different kind of call, one that was not familiar to me. Perhaps this species has a food-finding call, but more information is needed to establish this point. Many Ring-billed Gulls were attracted to the milling fiddler crabs, but they could have seen the other gulls foraging.

In the humid forests of the tropics a number of species of small birds are often seen foraging in small parties. These mixed-species flocks have recently been studied in some detail in the Panama Canal region. It was discovered that certain birds in the flock act as "nucleus" species to which the other kinds of birds in the group are attracted. The latter species may be regular or occasional associates in the flock. A number of these birds have been noted to be unusually noisy, uttering very loud call notes (or contact notes as they are often called). Probably the function of such loud and almost continuous calling is to keep the flock together, in contact with fellow members of the group. Furthermore, the loud and persistent calling serves to advertise the presence and location of the nucleus species to other kinds of birds. In this way it is easy for a number of birds to find the nucleus group, join it, and begin foraging as the group travels about. What advantage would accrue to birds if they showed such gregarious foraging behavior? No doubt it is easy for such birds to find food supplies in that more sense organs are available for food-seeking. If one species in the group finds some food, the others will be close by and can profit from the find. Similarly, if a predator is detected the other members of the group will be alerted to the presence of danger by hearing the first enemy-spotter call.

Speaking of danger, a most important communication function of many bird calls is to warn others of the presence of a potential predator. At the sight of a hawk or other kind of enemy, a small songbird will give its alarm call as it ducks for cover. (See Figure 45.) Others within earshot take heed and hide to protect themselves. The "meaning" of the alarm call is quite clear, even though the listeners do not see the enemy themselves. It is of interest to note that, like visual signals, alarm calls can be graded as well. For example, if you are walking near the edge of an active Green Heron nesting area and one of the breeding birds spots you, it will give a few rather low-pitched "skuk-skuk" calls. The bird that detected you is not particularly alarmed at the moment, and you may notice that its neighbors are not overly anxious, although they did look around and became aware of your presence. As you draw nearer to the alarmed bird, however, it will shift to louder alarm calls of the same type, and it will utter more of them per unit of time. Also, this bird may rise from its nest, and you can see that its neighbors are getting uneasy. If you come very close and force the bird from its nest, it will then shift to a very high-intensity alarm call of a different type: the bird stands upright, with its crest going up and down very rapidly, and it opens its mouth wide and utters a very rasping kind of call that spreads throughout the area and is easily detected by any heron within earshot. The bird first forced from its nest will fly from tree to tree around its own nest tree, and it peers intently at the intruder as it does so. It keeps up the alarm cries until the disturbance goes away. This kind of behavior is in marked contrast to what a Green Heron does when it spots an enemy such as the Marsh Hawk mentioned previously. Here, the heron is silent and adopts the motionless "bittern stance." In the case of the human observer the nest is in danger; hence, the loud alarm calls. In the second example, the heron is in danger; thus, it remains silent so as to be passed over by the hawk.

If we examine the alarm calls of a number of small songbirds by comparing sound spectrograms made from tape recordings of their voices, we will find that there is a remarkable similarity to them. When properly "read," such spectrograms tell us that the birds in question have alarm calls that are pure, high-pitched whistles or "thin," wiry squeeks. They tend to begin and end very imperceptibly. Such features as these are beautifully adapted to the situation in which such calls are normally uttered. A hawk suddenly appears on the scene and is spotted by a small songbird that utters an alarm call as it dives for cover. Other nearby birds hear the call and head for protective retreats themselves. Such an alarm call is very difficult to locate, which is an obvious advantage for the original caller. It does not want to be taken as prey by the hawk, yet there is an advantage for the species in alerting the others to the source of danger. Hence, the evolution of an alarm call that is easily heard by neighboring birds, yet at the same time is difficult to locate. The other small songbirds are not concerned about the location of the caller but only the alarm call itself, whereas the hawk is concerned with the location of the caller. This kind of alarm call is in marked contrast to the "mobbing" calls of these same small songbirds. (See Figure 45.)

Many kinds of small birds will show "mobbing" behavior if one or more of them chance to discover an owl resting in some bushes or trees during the daylight hours. At once the discoverer sets up a clamor and soon the owl is ringed by numerous small songbirds of a number of different species. All of them are giving their mobbing calls and milling about the hapless owl. The spectrograms of such mobbing calls are quite different when compared with the alarm calls of the same species. Mobbing calls have a very definite beginning and end, and they are very easy to locate. Again, this is clearly an advantage in that the first discoverer of the owl "wants" its position to be known so that it may be joined quickly by others in order to worry the owl. One mobbing scene that I witnessed in Knoxville, Tennessee, showed the speed with which recruitment may take place once an owl is detected. A mockingbird first spotted a Screech Owl, *Otus asio,* resting in some bushes near my room. At once it gave a mobbing call, and within two minutes the mockingbird was joined by two towhees, one cardinal, two sparrows and a second mockingbird. All of them flew about the owl, harassing it until the unfortunate bird flew away and sought peace elsewhere. My attention was brought to the scene by the initial mobbing call of the mockingbird. Small songbirds such as these quickly learn the meaning of each other's alarm and mobbing calls, and they respond accordingly.

Similar interspecific vocal communication systems are of great interest in the relationship between an African bird, the honey-guide, *Indicator indicator,* and several species of African mammals: man, baboons, and especially the ratel, or honey-badger, *Mellivora capensis.* Although the initial discovery of this bird's unusual habit was first reported in 1609, we owe most of what we know to the

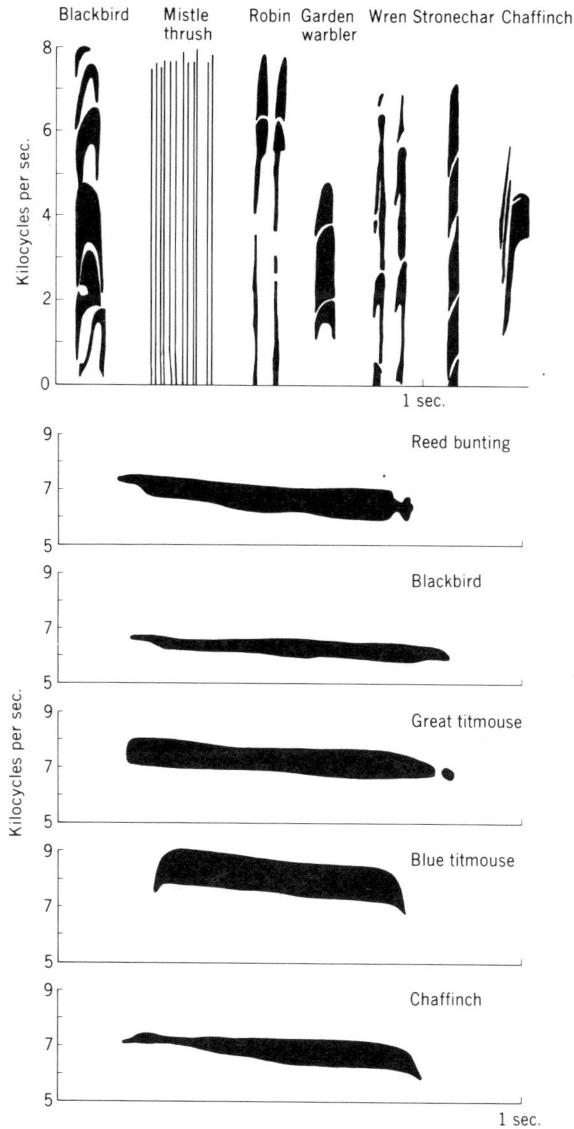

Figure 45. *Alarm calls of five different species of songbirds (below) are remarkably similar as shown in these sonagrams. By contrast, the "owl-mobbing" calls of seven different songbirds (above) are quite different. See text for details.*

thorough researches of Dr. Herbert Friedmann. Honey-guides are found in the Old World, mainly in Africa, and they are most famous for three characteristics of their way of life: 1) they are brood parasites, that is, the females lay their eggs in the nests of host species that give all parental care to the young honey-guides (a subject to which I'll return at length in Chapter VIII); 2) they eat wax (the only birds that regularly do so); and 3) they guide man, baboons and ratels to beehives. When a Greater Honey-guide wants to lead a follower to a hive, it makes itself very conspicuous by flashing its wings and tail in front of the intended follower. The honey-guide also brings attention to itself by making very loud, churring sounds. Natives of Africa who are aware of the bird's leading tendencies let themselves be led over the landscape by the flying bird. Eventually, the honey-guide stops near the nest of a certain bee species and then perches quietly in a nearby tree. The man (or ratel or baboon) that has followed its bird guide now proceeds to get the honey or other wanted thing from the hive and soon departs. The honey-guide now moves in and eats the larvae of the bee as well as bits and pieces of the honeycomb. The relationship between the honey-guides and their followers is of mutual advantage: both get something of value for their trouble.

The songs of birds seem most conspicuous to us in the spring of the year when they are busy with reproductive activities. It is no doubt true that the major communicatory functions of bird song have to do with the breeding cycle. Unless we are especially concerned with this aspect of bird biology, we might tend to dismiss bird song with a simple aside—it's loud or pleasant or what have you. But if we examine this feature of a bird's life in detail, we find that even a simple utterance has great potential for communicating information rapidly. All of us have no doubt heard the song of a male Red-winged Blackbird near some marshy area in the springtime. It is a relatively simple song often written as "ok-a-lee," but it can convey the following precise information: 1) species identity of the singer, I am a *Red-winged Blackbird;* 2) the sex of the singer, I am a *male* Red-winged Blackbird; 3) the exact location of the singer, I am a male Red-winged Blackbird *on my territory;* and 4) the mood of the singer, I am a male Red-winged Blackbird on my territory *who is ready to attack and chase intruders.* All of this information in what seems to be a simple "ok-a-lee" song heard in the spring!

Why should species identity be important in the marshland or anywhere else for that matter? Remember the case of the look-alike species. The females of such species may not be able to recognize males of their own kind by sight but only by their voices. In order for the proper matings to take place, it is essential that the right male and female do the mating. So, by having a specific, distinctive song, one not sung by any other species in the habitat, the correct matings are assured. We have used the simple case of a male Red-winged Blackbird singing in a marshland, but in forested areas of the world, tropical rain forests, for example, dozens of closely related species may be singing their advertising songs in the same place at about the same time. Thus, the problems of the various females of all of these species would be very much complicated if it were not for the specific distinctiveness of the songs of the males of their own species.

The sex of the singer is important in that his songs will signal to the females of his kind that a male of her species is singing in a certain place. Not only will she receive the signal that a male of her species is present in the same habitat, but she is also given information as to his exact location. The song will serve to locate the singer not only to females of his species but also to rival males. The latter are also warned by the singer not to trespass or else they will be challenged. In addition, the motivation of the male is conveyed by his song—he may be ready to challenge and fight, or to court a female.

Recall from our discussion of animal speciation that newly evolved species, or at least forms that we suspect are on their way to becoming full-fledged species, must be reproductively isolated in order to maintain their new status. Sharp divergence in their advertising songs is the rule so that no mistakes are made by prospective mates. Furthermore, analysis of the song repertoire of many songbirds indicates that there is much individual variation in the song. Such variability could be the basis for individual recognition, not just that the singer is a male of a certain species. It has been shown in some species of birds that pairs that had been mated in a previous year or years were more successful in rearing their offspring. Thus, recognition of a mate from a previous breeding season could confer an advantage on a female that hears her former mate singing when she is ready for reproduction. Also, since a great deal of behavior in the early stages of pair formation is a matter of appeasement, individual recognition of a former mate could conceivably enhance this aspect of the breeding cycle.

What about young birds in the nest during periods of singing by one or both of their parents? This raises the question of whether or not the characteristic song of a species, and its many variations, are learned from the parents or innate. It is a fair generalization to state that the call notes of a species tend to be inherited. That is, the alarm notes, hunger calls and the like have an inborn basis. With respect to the full song of a species, however, there is much variability from one species to another. Doves, for example, apparently cannot modify their basic song by learning processes. This is not true, however, of many other kinds of birds. A great deal of recent work on the European Chaffinch, *Fringilla coelebs,* is of great interest in this regard. A common technique used by students of bird song and its heredity is to rear very young birds in complete isolation from others of their kind. They are isolated from all sources of sound except those that they produce themselves. When this is done with young chaffinches, it has been discovered that when they sing for the first time the song is of a rather simple nature. It does not closely resemble the "normal" song of the species as ordinarily determined. Now, however, if a continuous loop tape of a real chaffinch song is played back to such young, isolated birds, they will pick up bits of the taped version and incorporate them into their own songs. The end result is a song that somewhat approaches the "typical" one. If several of these young and isolated chaffinches are now allowed to group together, they will learn from each other (aberrations and all) and soon they will be singing a group song of their own concoction, even though once again it does not agree with the song characteristic of the chaffinch.

If a young chaffinch is trapped alive in the wild in the first autumn of its life and is observed for its various song characteristics the next spring, it is heard to sing a song like that typical of the species. This suggests that the young chaffinch has had the opportunity to hear other chaffinches sing while it is still in the nest and for a short time afterwards. Observations in the field have confirmed this assumption. Having heard typical chaffinch songs, the young bird will sing the song of its species. In fact, in its next spring the young chaffinch will sing about five or six themes that agree very closely with the local "dialect" in the area in which it was hatched; furthermore, it will add some individual notes of its own. Thereafter, such a chaffinch is heard to sing this kind of song season after season. If such a bird is trapped and then attempts are made to teach it new phrases and so forth, the attempts always fail. If, however, a bird reared in isolation is exposed to such "foreign" sounds, it will add bits and pieces of these songs to its own repertoire, but only if the foreign songs resemble somewhat the songs of a chaffinch.

It might be reasoned that the bird stimulates itself, and this element could be important in the development of its characteristic song patterning. Experiments with deafened European blackbirds suggests strongly that such self-instruction is not important in this species, because the birds sang the full song characteristic of their own kind. This, however, may not be true for all songbirds.

Why have vocal communication if you already have a full range of visual displays with which to communicate? Sounds travel far and fast, whereas visual signals must usually be seen close up for maximum effect, although there are some notable exceptions. Sounds are easy to produce, relatively little energy is used in their production, and they decay fast; that is, as soon as the sound dies out, another can be produced for a rapid series of vocal signals. If a visual displayer is hidden from the view of the receiver, the communication will be blocked. Sound by contrast is relatively unhampered by physical objects in the environment of the songster. Thus, communication may still be effected. A large range of intensities and frequencies is available to bring about a very complex array of vocal signals. Recall that the hawk alarm calls of small songbirds are notoriously difficult for the bird of prey to localize. By contrast, a flashy visual signal, if used to warn others of the presence of danger, would render the signaler vulnerable to attack because of the ease with which the display could be located.

As with so many things biological, we find that several

systems are employed, each for maximum effectiveness in certain environments. Some visual signals are used for long-range purposes of communication, while others require that the sender and the receiver be very close to one another. One signal may communicate general information, such as "I am here," while another may be highly specific, such as "don't touch me because I'm dangerous or taste bad," and so on. The habitat of an organism may determine in great part what signal systems are paramount. For example, birds that live in dense cover may rely heavily on sound communication because the thick vegetation of their habitat reduces the effectiveness of a visual signal. A bird of the open grasslands may sing while in flight to further enhance its displays. Recent work on bird species that sing "duets" has been most rewarding in this regard. A bird from Africa called the Boubou Shrike, *Laniarius aethiopicus,* widely distributed in that continent, has been studied in detail from the viewpoint of its duetting vocalizations. This species live in pairs in quite dense cover. Visual signaling is more or less inefficient in keeping the pair in contact for obvious reasons. Thus, there has evolved in this shrike an elaborate system of what has been called "antiphonal" singing. That is, one member of the pair sings a song, and this is at once followed by the song of the other member of the pair. The timing, the synchrony, of the two sets of songs is so precise that the observer believes that he has heard one individual sing a rather complex song. Either sex can initiate the duet, and if one member of the pair is absent from the territory momentarily, the remaining mate sings the songs of the other bird. In fact, one of the functions suggested for such duetting is to get the "missing" mate to return! It is reasoned that the bird away from the territory hears its own song being sung, and then returns at once to see if possibly an intruder is trespassing on the territory of the pair. In this way, a very precise method of keeping track of one another has evolved in this most fascinating African shrike.

Ground- and Cliff-nesting Gulls Compared

Habitat can influence behavior in many ways. We have just seen one example—the influence of dense vegetation on signal system evolution. Another fine example comes from the comparison of behavior and ecology of gulls that typically nest on the ground and those species that have taken to nesting on cliffs. (See Figure 46.) Let's briefly examine the behavior of a "typical" ground-nesting gull such as the Herring Gull. Recall that in this species the male is very agressive and threatens rival males regularly. It stakes out a territory for itself, advertises its presence on the defended area, engages in grass-pulling activities with its rivals, and so forth. One of the more striking hostile postures of this species is the upright threat and the stiff, slow walk towards the opponent. Furthermore, alarm calls are sounded regularly and loudly by the Herring Gull whenever danger draws near. The chicks of this species are protectively colored, and they crouch on the ground and remain immobile whenever the alarm call is sounded by the adults. This is clearly related to a rather high danger from predators in a crowded ground gullery.

By contrast, the cliff-nesting gulls, especially the Kittiwake, *Larus tridactyla,* show some marked deviations in their nesting behavior when these attributes are compared with those of ground-nesting gulls. Since predators find it very difficult to gain access to the nests of this species located on tiny ledges on a steep seacliff face, we find that the Kittiwake very rarely utters an alarm call. In addition, the chicks of this gull are not at all cryptically colored, nor

Figure 46. *Ground-nesting gulls have separate nest defense displays (above right) and advertising displays, including territorial proclamation by vocalization (below right); by contrast, the Kittiwake (left) combines both in one display as it stands on its narrow cliff ledge territory.*

do they show any crouching behavior when they are alarmed. Rather than having an elaborate series of upright hostile postures, the Kittiwake engages mostly in beak-grabbing aggressive behavior; that is, Kittiwake rivals attempt to twist the head of their opponent and toss it over the cliff. The upright postures of the ground-nesting gulls could not function properly on the narrow ledges of the Kittiwake's habitat. Also, a threatening Kittiwake can't walk very far in an intimidating posture because it would very quickly trespass on a neighbor's territory. The crowded nest sites on the cliff ledges result in very small territories for each Kittiwake pair.

As a further adaptation to life on the cliff, adult Kittiwakes build a very elaborate nest, one with a heavy mud cup as the base. This is essential so that the eggs will be contained in one place and not roll over the edge of the cliff. Ground-nesting gulls, by contrast, build very simple nests consisting of a few bits of vegetation, debris, and so forth. There is no mud cup. Again, this is an adaptation for nesting on the ground. Kittiwakes also tend to synchronize their gathering of nesting materials. Flocks leave the cliffs and gather mud and other materials so that we find a large number of Kittiwakes at the same stage in the nesting cycle. On the other hand, ground-nesting gulls do not show social gathering of nest material, and we find relatively little synchrony to nest-building.

When the young Kittiwakes hatch, they spend most of their time facing the wall of the cliff to prevent their being blown over the side by the turbulent winds whipping around the face of the cliff. In addition, the young have well-developed claws to help them in holding on to the nest tightly. As they grow larger, the young Kittiwakes do not engage in very much wing-flapping, a behavior that might eliminate them rapidly in their highly specialized nesting environment. On the other hand, young ground-nesting gulls move about a great deal, they do not have well-developed claws, and they engage in a great deal of wing-flapping before they become independent.

Thus, nesting on a tiny cliff ledge has brought about many changes in the biology of the Kittiwake: structural, behavioral and ecological. Such changes came about slowly, through the mechanism of natural selection working on Kittiwake variability. Any variation that promoted success in the cliff-nesting way of life would be favored, while variants that did not confer an advantage would be eliminated in this stressful nesting environment. Gradually, over long stretches of time, with the coming and going of many Kittiwake ancestors, adaptations such as the ones we see in the Kittiwakes of today would have accrued to the species. Recent studies of cliff-nesting gulls in the high Arctic also indicate how such patterns of behavior and ecology in living Kittiwakes may have evolved in the past. These cold-country relatives of the Kittiwakes seem to be almost perfectly intermediate between the typical ground-nesting species as exemplified by the Herring Gull and the very specialized cliff-nester, the Kittiwake, inhabitant of the lonely sea cliffs.

VIII

Brood Parasitism

It is fascinating to think about such an unusual aspect of bird biology as brood parasitism. You will recall that this refers to the behavior of some species of birds in which the females lay their eggs in another's nest so that all parental care—incubation of the eggs and feeding of the parasite's young—devolves on the host. A dozen questions spring to mind: how did such a pattern of behavior originate and evolve? Is it widespread in the world of birds? Do the hosts have any "counterattack"? Have any unusual specializations of structure and/or behavior evolved in the parasites in conjunction with such activities on their part? And so on.

Occurrence in the Bird World

Brood parasitism is known to exist for five families of birds, although a sixth may sometimes be listed. Some professional ornithologists interested in matters of classification choose to "split" one of the families into two separate ones. Here I am considering only those bird species that are regularly brood parasitic; that is, they are obliged to be so as a normal part of their lives. I am not going to consider those birds that appropriate the nest of another species, or those that will, on occasion, lay an egg or two in the nest of a different species. However, sometimes clues as to the origin of brood parasitism, as we shall see, can come from close observation of those birds that take over the nest of another species for their own purposes, or that do lay an egg now and then in the nest of a "host."

The bird family Anatidae, which includes the ducks, the geese and the swans, has one member that is a brood parasite. This is the Black-headed Duck, *Heteronetta atricapilla,* of South America. The Cuckoo Finch, *Anomalospiza imperbis,* a species that has a very wide distribution in Africa, is classified in the family Ploceidae; it, too, is a brood parasite. The eight species of combassous and widow birds are sometimes placed in the same family, Ploceidae, as the Cuckoo Finch, or else they are classified in the family Estrildidae. These combassous and widow birds are all brood parasitic, and they have very broad distributional ranges throughout much of the African continent. One species occurs on several islands in the Atlantic Ocean off the west coast of that continent as well as on several islands off the east coast, out in the Indian Ocean. Two additional Old World families that contain brood parasitic species are the Cuculidae, the great family of cuckoos with a great many brood parasitic forms, and the family Indicatoridae, the honey-guides, already mentioned in another context. The only New World songbird family with brood parasitic species is the large family of blackbirds—the Icteridae. This family includes such familiar members (to North American bird observers) as the Red-winged Blackbird and the Yellow-headed Blackbird, mentioned previously in the chapter on territorial behavior, as well as meadowlarks, orioles and grackles. Less familiar are such Central and South American species as caciques, oropendolas, and so forth. One group of the blackbird family, the cowbirds, are all brood parasites with one exception. This latter species is most interesting in that careful study of its breeding biology has provided us with a number of insights in to how this unusual mode of reproduction in the cowbirds may have originated and evolved.

Adaptations for Brood Parasitism

Before we look in detail at the various species of brood parasites, we might ask at the outset what adaptations for this mode of life have evolved in the different species that show these patterns of letting others rear their young. First, a problem for the female of a brood parasitic species is to find the nest of a suitable host, a nest that is at the right stage in the breeding cycle. The nests of some species that might act as potential hosts may be difficult to locate; they may be placed in rather inaccessible places such as holes in trees, or banks of streams, where the kingfishers nest, or else in tangled, dense undergrowth. Perhaps the nest, when found, will not be at the right stage in the cycle. If a suitable nest at the right stage is found, what is the incubation period of the host's eggs? If the host warms its eggs for twelve days and the brood parasite were to require fifteen days for successful incubation, then the tiny brood parasites upon hatching would be at a clear disadvantage. The brood parasite's nestmates would have the jump on it in the vital matter of getting food from the host parents. Chances are, such a brood parasite would succumb by starvation. So, matters of nest finding, stage in the breeding cycle of the host, and the incubation period of the host's eggs are of vital concern to the welfare of the brood parasite's eggs and young.

Furthermore, what kinds of foods does the host species feed its own young? If the food needed by the hatchling brood parasite is quite different as compared to the typical diet fed to the host's young, then the little brood parasites would be in great difficulty. What about the appearance of the brood parasite's eggs and young? A simple experiment such as the following has been tried many times in the field: an odd-looking egg is placed in the nest of a small bird; that nest contains several of the bird's own eggs. Upon returning after the observer has hidden, the small bird is seen, in many species, to eject the odd-looking egg. In fact, if the observer now removes all of the eggs of the host and replaces that set with one containing several parasite's eggs and one of the host's, the small bird will return and eject its own egg! In other words, for many species of small birds, it is the odd object in the nest that is tossed out. Under normal circumstances, of course, the odd egg would be that of some brood parasite. Suppose the brood parasite were to lay more than one egg in the nest of some host? Would the host adults be able to feed more than one young brood parasite if they accepted the eggs and incubated them to the hatching stage? What would happen if the young brood parasite did not respond exactly like the young of the host species? Would the host feed such a youngster, or toss it out, or kill it? Suppose the nestling plumage of the parasite's young did not match that of the host's own young? Would the intruders be fed? Would they be killed in the nest, or simply tossed out, or what? These and numerous other questions can be asked about this most interesting facet of bird reproductive biology. We have answers, or partial answers, for some of these questions, but it can be honestly said that we are just beginning to find out the details of life in most brood parasitic species. Most of the world's brood parasites are known but sketchily. A number of men have labored to answer these and related questions on the biology of brood parasitism, but it is fair to state that we owe most of our current knowledge of this kind of reproduction to the life-long efforts of one man, Dr. Herbert Friedmann, formerly of the United States National Museum in Washington, D.C., but now Director of the Los Angeles County Museum.

What are some of the more striking adaptations of brood parasites? First of all, with respect to the matching of eggs between parasite and host, we find that the parasitic cuckoos have evolved much more detailed egg mimicry than other kinds of brood parasites. In fact, it may be extremely difficult for a human observer to tell the cuckoo's eggs from those of its primary hosts. Presumably, such an adaptation has evolved to counter the response of the host to an odd object in its nest; that is, to prevent the host from tossing the parasite's egg over the side of the nest. In addition, since many of the cuckoo's hosts are smaller than it in body size, the eggs of the brood parasite tend to become smaller. For example, the Yellow-billed Cuckoo, *Coccyzus americanus,* which is a non-parasitic species, is smaller in overall body size than its relative the parasitic European Cuckoo, *Cuculus canorus,* yet the American species lays larger eggs.

If the behavior of an individual female European Cuckoo is examined, we find that she tends to lay her eggs in the nests of only one host species, that is, she tends to be host specific but in an emergency she will lay in other host's nests. It is generally assumed that this primary host is the same species that raised this female when she was a nestling; however, this aspect of brood parasitism is very poorly known. For example, one European researcher studied the behavior of a single female over several breeding seasons, and he found that this one female European Cuckoo laid 61 eggs over four different breeding seasons. Of this total of eggs, 58 were laid in the nests of but one host species, the Meadow Pipit, *Anthus pratensis.*

If female cuckoos tend to lay their eggs in the nests of but one host species, then we should find that several females in one area are parasitizing different hosts. Such an observation was noted a number of years ago by the famous British ornithologist Alfred Newton; in fact, Newton concluded that a cuckoo species was in reality divided up into smaller groups which he called "gentes." The female members of the same gens (singular of gentes) all lay their eggs in the nests of the same species of host. Such eggs are similar in size and coloration and match those of the common host. Members of another gens will lay eggs of a different size and coloration to match the eggs of their own prime host, and so forth. For example, in Finland it is possible to find three main types of cuckoo eggs: 1) blue, unspotted eggs, with the chief victim of members of this gens being the European Redstart, *Phoenicurus phoenicurus;* 2) pale blue eggs with heavy reddish spots on them, the chief host of this gens being the Brambling, *Fringilla montifringilla;* and 3) eggs with a white background and fine gray markings all over the shell, with the chief host being birds called wagtails, *Motacilla alba.* This matter of gentes poses a very difficult problem, however: a male cuckoo with a very large territory may mate with several females, each of which may belong to a different gens. It would seem that the niceties of this fine egg mimicry would be lost because of the genetic contribution of one male to all of his mates. One suggested explanation is that perhaps those genes involved in the exact egg mimicry are in fact located on the female's odd sex chromosome. In birds, unlike man and other mammals, it is the female that has the unlike pair of sex chromosomes. At any rate, this aspect of brood parasitism provides some stimulating food for future thought and much hard research in the field before anything more definite can be stated.

How does a female brood parasite find a suitable host or at least its nest? Perhaps the female simply looks at random in her overall search area, and lays her eggs in the nests of suitable hosts thus discovered more or less by chance. Or, the female might watch the nest-building activities of her prime host very carefully, and then she would quickly move in and deposit her eggs at the proper time. Some observations made in Germany and Finland suggest that the truth may lie somewhere in between. Probably the female cuckoo takes advantage of the alarm behavior of her hosts; in other words, the host gives away the location of the nest to the brood parasite by giving its alarm call when the cuckoo comes near. Small songbirds will give varying degrees of alarm when a potential predator approaches the nest: low-level when the danger is still far away, higher intensity when the nest is about to be discovered. Thus, the female cuckoo receives both visual and auditory clues as to the exact location of the nest by the activities of the host. It must be said, however, that this aspect of brood parasitism is very poorly known.

Once the eggs of the parasite are in the host's nest, are further parasite adaptations necessary? The answer is an emphatic yes! The egg or eggs must hatch in about the same time as the eggs of the host; thus, we find a shortening of the incubation period for the eggs of some kinds of brood parasitic birds. Additionally, the young brood parasite must compete for food and also be able to utilize the foods brought back by the foster parents. The young of the European Cuckoo have a most interesting behavioral response to the presence of other objects in the nest. When it is less than one day old, the little, unfeathered cuckoo, with its eyes still closed, gropes its way about the nest.

At this time it has a rather shallow pit or depression in its back that is sensitive to objects with which the little cuckoo comes into contact. If such an object happens to be an egg or a nestling of the host, then the little parasite wriggles and squirms its way under the object until the object lies in the hollow. Then the cuckoo continues to wriggle and squirm its way up to the edge of the nest and heaves the egg or nestling over the side! Thus, the young cuckoo ensures itself of undivided attention from its foster parents. The young of several species of African brood parasitic honey-guides come equipped with even more deadly devices: they have a pair of sharp-pointed hooklets on the tips of their mandibles. When they come in contact with another young bird in the nest, they struggle fiercely and pierce the flesh of the host's young until the other birds are dead. Again, this is an adaptation that ensures parental care via the elimination of host-young competition.

In the combassous and widow birds of Africa, we find that the young have yet another device for duping the foster parents. These nestlings have colored spots on the inside of their mouths as well as globular spots at the corners of their open mouths. The inside spots contrast nicely with the general background coloration of the mouth lining, while the globular tubercles are highly reflective and seem to glow in the dark. It was thought at one time that such tubercles were phosphorescent; that is, it was believed that they had their own light-emitting ability, but more recent studies have shown that the tubercles are reflective in nature. Combassous parasitize very close relatives of theirs, ones that nest in globe-shaped structures with dark interiors. Hence, reflective structures will show up like so many tiny light bulbs in the dim light of the nest's interior. In this way, the tubercles act as feeding guides for the foster parents entering the nest to feed the brood inside. The mouth and tubercles of the young parasites resemble closely those of the host's young. However, it should be pointed out that students of the classification of these birds feel that the similarity of these devices between the parasites and their hosts has not evolved through the agency of brood parasitism but rather reflects the very close relationship of the two groups of birds. In fact, close resemblance of the mouth markings due to common ancestry may have been an adaptation favoring the success of the brood parasite rather than a result of this mode of reproduction.

Not all brood parasites have egg mimicry, plumage matching between nestling parasite and host young, mouth markings, mandibular hooks, hollow backs for ejection of host young, and so forth. Each species has its own concatenation of attributes that render it more or less adapted for the life of a parasite. The females of some species remove a host's egg before depositing one of their own. Some destroy the entire clutch of the host. The glossy cuckoos of Africa have even been observed feeding their own young *after* they had been brought to the fledgling stage by the host adults! Thus, generalizations about brood parasitism must be carefully considered because there are so many exceptions. Let's turn now to a detailed consideration of the brood parasitic way of life among some widely different species that practice this kind of reproduction.

A Brood Parasitic Duck

The Black-headed Duck is a bird of dense marsh vegetation, and it is found in central and southern South America. Records of species that it has been known to parasitize include other species of ducks, ibises, coots, hawks, screamers, herons, storks and gulls. Although its parasitic way of life has been known or suspected for decades, little detailed information was available until the fine studies of Milton Weller in Argentina in the mid-1960's. Weller concluded that this duck is undoubtedly completely brood parasitic since no nests of the species were ever found; in addition, no females were ever seen with broods of ducklings, and no brood patch was ever found on the females. More importantly, Weller was able to add considerably to our previously meager knowledge of this bird. For example, he found that the principal hosts in his study area in Argentina were the Red-fronted Coot, *Fulica rufifrons*, the Red-gartered Coot, *Fulica armillata*, the Rosybill Duck, *Netta peposaca*, and the White-faced Ibis, *Plegadis falcinellus*, with the Red-fronted Coot being the most important host species.

All of the hosts nested in the dense marshes of the study area, but it was difficult for the observers to see the actual mode of parasitism; hence, many details are still lacking on the biology of parasitism in this duck. However, it was determined that the female Black-headed Duck does not destroy nor disturb the eggs of the host species. She does not destroy or disturb the nest, and she tends to lay but one egg in any one host's nest. However, the number of eggs in one host nest might range from one to a maximum of eight. Most interesting was the behavior of the young duckling parasite upon hatching. This duckling remains in the nest of its host for about one to one and a half days; then it leaves the host's nest and becomes completely independent. Apparently, the duckling leads a completely solitary existence, for no gatherings of young ducks of this species were ever seen on the study areas, even though intensive searches were made for them by Dr. Weller and his associates. The young ducks rarely gave alarm calls, they were able to care for their plumage almost from the moment of hatching, and they showed excellent foraging behavior right from the start. All of these adaptations would clearly aid in the solitary life of the newly independent duckling. No food at all is provided the young duck by its foster parents, and in this respect the Black-headed Duck may be considered a most successful brood parasite in that no damage whatever is done to the hosts by either the parent brood parasites or their young. Additionally, no specializations, such as egg mimicry, adaptations for removal of the host's eggs or young, and so on are known for this most interesting relative of the swans and geese.

Widow Birds and Honey-Guides

The Paradise Widow Bird, *Steganura paradisaea,* is found commonly over a wide area in the more open parts of Africa, and this species is a popular cage bird in captivity. The male in breeding plumage is a very showy individual with blackish head, gold-colored nape, chestnut chest, and a long, most conspicuous black tail. It is the latter structure that really sets off the male of this species. In his courtship display, the male flies in a jerky fashion from one treetop to another, and then he spirals high in the sky, perhaps up to three hundred feet, with his long black tail still making him conspicuous. Now, he plummets towards the watching females perched far below him. It is not known for certain, but strongly suspected, that the male widow bird will mate with more than one female. The eggs are white and tend to be placed in the nests of closely related weaver bird species. However, it is rather remarkable that few details are known about the brood parasitism of this species because it is a common and very conspicuous bird. Host species usually cited are the Melba Finch and its close relatives the Yellow-winged and Yellow-backed Pytilias.

More is known, by contrast, of the parasitic way of life of the honey-guides, the indicator birds of Africa, largely through the efforts of Dr. Herbert Friedmann, mentioned earlier. Honey-guides are related to woodpeckers, barbets and similar hole-nesting or burrowing birds, and it is among such species that we find the most common hosts of the honey-guides. Woodpeckers and their allies tend to have relatively short incubation periods but rather longer nestling growth periods. The honey-guides show similar trends. However, at least one honey-guide parasitizes hosts that build open nests and have quite short nestling stages; nevertheless, the young honey-guides are reared by the foster parents. Honey-guides lay plain white eggs as do their prime hosts, the woodpeckers, barbets and similar birds. This appears to be an adaptation for nesting in burrows, cavities, holes in trees and the like where color and patterning would be much less useful than in open-nesting situations. Even when honey-guides parasitize hosts with colored eggs, however, they seem to be just as successful. Thus, close egg mimicry does not seem to be "required" for success, at least in honey-guides and at this stage in our knowledge of this group of birds. The most remarkable feature of the brood parasitic way of life in these birds is the presence of mandibular hooks on the bill of the young honey-guides, mentioned previously in our consideration of specializations for the brood parasitic way of life. With respect to other features of honey-guide biology, we find that they are also remarkable for their guiding habits (recall the African ratel being led to a beehive) and their ability to eat and digest wax. Brood parasitism is strange enough, but the honey-guides offer numerous other features ripe for further research as well.

The Cowbirds

Turning now to the New World, we find that the cowbirds of the large blackbird family have among their number some brood parasites; study of these birds have revealed

—in part at least—how this pattern of reproduction may have originated and evolved. The parasitic species are the Shiny Cowbird, *Molothrus bonariensis,* the Screaming Cowbird, *Molothrus rufo-axillaris,* the Brown-headed Cowbird, *Molothrus ater,* the Bronzed Cowbird, *Tangavius aeneus,* and the Giant Cowbird, *Scaphidura oryzivora.* An additional species must be mentioned even though it is *not* a brood parasite; this is *Molothrus badius,* the Bay-winged Cowbird. Although non-parasitic, this species is important in helping to unravel the complicated story of the origin and evolution of this fascinating behavior.

Brood parasitism in the cowbirds ranges from extreme host specificity to unusually broad parasitism involving over two hundred different kinds of hosts. The Brown-headed Cowbird is the most broadly parasitic of the cowbirds, but the over two hundred records include a number that may be doubtful or only occur infrequently. Furthermore, the fact that a cowbird female lays an egg or more in a host's nest does not insure that the young cowbird will reach the fledgling stage. Most of the records of Brown-headed Cowbird parasitism, in fact, come from but fifty species of hosts. Some, such as the Song Sparrow, the Yellow Warbler, and the Red-eyed Vireo, are much more common victims of the Brown-headed Cowbird.

The cowbirds are an extremely successful group, but it is of interest to note that they lack the adaptations for the brood parasitic way of life found in the previously discussed cuckoos, widow birds, and so forth. For example, egg mimicry tends to be lacking, except for the Giant Cowbird of Central and South America. There is no increase or decrease in the size of cowbird eggs in relation to the egg size of their hosts. Young cowbirds do not kill nor do they evict the young of the host from the nest containing both of them. Laying females may remove the eggs of the host from a nest about to be parasitized, it is true, and the foster parents of the young cowbirds may give them the larger share of food—to the detriment of their own young. But the highly specialized structures and behavior of the honey-guides and the cuckoos, for example, are completely lacking in the cowbirds, yet this way of life has been eminently successful for them.

What about the origin and subsequent evolution of this kind of behavior in cowbirds? Looking at the only cowbird that is not a parasite, the Bay-winged Cowbird, we note that this species uses other birds' nests for its own purposes, but it always incubates its own eggs and rears its own young. At times, this species even tries to build its own nest, but often without success. A possible next step in the evolution of brood parasitism might be for a bird to lay its eggs in the readily available nests of other species, but still build its own nest once in awhile. This stage is actually shown today by the Shiny Cowbird. Intermediate steps might be the reduction in number of hosts parasitized, with further loss of the nest-building habit. Finally, the culmination of this behavior might be extreme host specificity and complete loss of any tendencies to build a nest. This stage—parasitism of but one host species—is shown today by the Screaming Cowbird, and it appears to be a quirk of nature that the only host of this, the most host-specific of all of the cowbirds—is the only cowbird that is not a brood parasite, the Bay-winged Cowbird! This may not be the actual end point of specialization for the cowbirds, however, but may be an accident of the timing of the breeding seasons of these two cowbird species in South America. Both the Bay-winged and the Screaming Cowbirds breed late in the South American breeding season; hence, a number of potential hosts for the Screaming Cowbird are not available for parasitism since they have completed their nesting cycles. But, the Bay-winged Cowbird is handy for parasitism. A likely next step for further specialization of the brood parasitic way of life would be the evolution of structural and behavioral changes further enhancing this mode of reproduction. For example, egg mimicry, behavior of the young cowbirds, changes in the behavior of laying females, and so forth. As mentioned above, however, the cowbirds have done rather well so far without unduly specialized structures and/or behavior.

It is usually considered to be a truism that brood parasitism is detrimental to the host and beneficial only to the young brood parasites. For example, the actual destruction of eggs and nestlings by young cuckoos and honey-guides with their specialized structures and behavior, or the puncturing and tossing out of host eggs by the laying female brood parasite. Young parasites may also outstrip the host young in development by taking the lion's share of food brought by the host adults. But is brood parasitism always detrimental to the host species? A recent study in Panama suggests very strongly that this way of life can have advantages for the host species as well as for the brood parasite.

The Case of the Giant Cowbird

This case of parasitism with advantages on both sides concerns the Giant Cowbird of Central and South America (see Figure 47), which is largely parasitic on other members of the blackbird family, most notably the colonial nesting oropendolas (see Figure 48) and caciques. Unlike the other parasitic cowbirds, females of the Giant Cowbird species may lay eggs that mimic those of their hosts. Studies showed that there were in reality five kinds of Giant Cowbird females with respect to the eggs laid by them: three kinds laid eggs that mimicked the eggs of three kinds of oropendolas that were being parasitized. One

Figure 47. Giant Cowbird (Scaphidura oryzivora) *of Central and South America is a brood parasite.*

kind of female laid eggs that closely mimicked the eggs of its host, a cacique. Still a third laid a non-mimetic egg in the nests of its hosts. In addition to the differences in the kinds of eggs they laid, the female Giant Cowbirds also differed strikingly in their behavior before they laid an egg in a host's nest. Those females that laid matching eggs were quite shy and moved cautiously through the colony of their hosts; such females were waiting patiently for their hosts to leave their nests so that they could enter and lay an egg. By contrast, the females that laid non-mimetic eggs were very aggressive and actually drove off the hosts to parasitize their nests.

Several additional unusual findings related to the ability of the hosts to tell Giant Cowbird eggs from their own. Some caciques and oropendolas could readily discriminate

Figure 48. Wagler's Oropendola (Zarhynchus wagleri) *is often parasitized by the Giant Cowbird, but not always to its detriment (see text for details).*

between the parasite's eggs and their own. Their reaction was to toss out the foreign egg or eggs. Other individuals could not make such a distinction, but it is of great interest to note that the ability to discriminate or not was a *colony* attribute. That is, all of the caciques or oropendolas in a colony either discriminated or not. These studies also showed that some colonies of the host species reared an unusually high number of the young parasites, while other colonies of equal size did not rear as many young Giant Cowbirds. As you might expect, those colonies that were composed of non-discriminators with respect to egg-matching were the ones that reared large numbers of the young cowbirds. Related to these features was a behavioral difference in males of the cacique and oropendola species: males from colonies that rejected many of the cowbird eggs were aggressive and chased off the Giant Cowbirds, while males from those colonies that accepted many of the cowbird eggs did not chase the parasites from their nests.

What about the success of the parasitized nests? This, to me, was one of the most interesting findings of this remarkable study made by Dr. Neal G. Smith. Nests of oropendolas and caciques may be placed very close to, or even around, the nests of wasps and stingless bees. Although the latter bees are stingless, they do bite. This would clearly appear to afford protection from various predators. Or, the nests of these colonial blackbirds may be clustered at the tips of parasol-shaped trees, making it very difficult for some terrestrial predator to gain access to the colony.

More importantly, however, the chicks of the caciques and oropendolas probably suffer more losses from the parasitism of botflies. The female botflies lay their eggs or their living larvae in the skin of the young caciques and oropendolas. If a large number of the larvae thus deposited on a chick thrive, then that chick usually succumbs from the infestation. However, in those nests of the caciques and oropendolas that are parasitized by Giant Cowbirds, the losses of the host young are very greatly diminished because of the behavior of the young parasitic birds. The nestling Giant Cowbirds preen the nestlings of the caciques or oropendolas. Thus, they remove the eggs, larvae and even the adult botflies from their skin. As a result, more young hosts are reared from nests with young cowbirds in them than from non-parasitized nests!

In summary, we know a great deal about the brood parasitic way of life in some species, but as Dr. Smith's studies have revealed, much is to be learned about the behavior of many more species of these unusual birds. The origin and evolution of this behavior is not well understood, although some tentative hypotheses have been put forth. What brings about loss of the nest-building habit in these birds? What about those species that occasionally "dump" eggs into another species' nest because of lack of synchrony between nest-building and egg-laying behavior? Does this have any bearing on the origin of brood parasitism? We have a great deal of information on hand, we have a number of ideas to mull over, but much remains to be done in this most fascinating area of bird behavior.

Epilogue

Birds as we have seen are suitable organisms for research on a variety of biological problems. I have emphasized their use in evolutionary and behavioral work, but a number of biological sub-disciplines have gained much information from the observation and experimentation done with birds. Have these many and varied researches, some of which I've discussed, exhausted the use of birds in our attempts to expand our knowledge in the areas of evolution and behavior? I would say that we are at the beginning rather than the end point of man's use of birds as research tools in these two disciplines (and others in biology).

What of the future? With each new breakthrough in a different discipline of biology, the techniques discovered and employed can be used in studies of evolution and behavior. For example, the past decade has seen much progress made in biochemical analyses. Recent proposals for revision of our views on bird classification, to cite just one example, are based heavily on the determination of various body proteins and how they may be related in different species and higher groupings of birds. Similarly, a relatively recent innovation is known as numerical taxonomy, and many new proposals for animal classification, based on the use of this technique, are appearing in the scientific journals at an increasing rate. The "bandwagon effect" is a truism in science as well as in the arts!

It was once said, perhaps with some measure of truth, that ornithologists stuffed and saved the skins of birds but threw away the best parts. Today, new ways of looking at the fine structures of feathers, new techniques for the analysis of pigments, and so forth, suggest that the bird researchers are re-examining their material to avoid this old criticism. Additionally, detailed studies of blood pigments and related biochemical endeavors are adding greatly to our fund of general knowledge as well as helping us to reassess our views on bird evolution and classification.

The simple question, "What is a flamingo?" is easier to ask than to answer. Some workers believe that these elegant, long-legged waders are more closely related to storks and herons than they are to other bird groups; by contrast, some workers feel that flamingos are really odd kinds of goose-like birds, while still other students feel that these birds are so different that they warrant a classification that reflects their distant relationship to other kinds of birds. Each group has made use of information gathered from studies on the structure, behavior, biochemistry and parasites of flamingos, but the results are obviously in conflict. Perhaps additional biochemical or behavioral information, or re-evaluation of previous data, may help us solve an old ornithological problem. Sometimes a new tool, such as the electron microscope or the magnetic tape recorder, will provide the needed data for our judgments. Sometimes, however, the results are not decisive and we must await the advent of a new tool, or a new technique, or a new mind to look over the existing information and interpret it in a new way.

With respect to behavioral studies, Professor Niko Tinbergen of Oxford University has stated, ". . . simple, precise observation . . . in the young science of ethology . . . is of supreme importance. . . ." This should be our guide. It is true that many sciences rely heavily on masses of very expensive equipment for their survival, and it is true that ethologists have and do use expensive recording equipment such as cameras and tape recorders, but it is also true that keen eyes, patience, paper and pencil have and will make notable contributions to this most exciting field.

Furthermore, advanced degrees in ethology are not necessary in order to make important contributions to this area of animal behavior. Patience and careful descriptions of behavior are the frameworks. Later, experimentation—often done right in the field—will enable the observer to test—with simple apparatus—his preliminary ideas gained from prior observations. Again, Professor Tinbergen clearly states: ". . . the natural environment is an extremely important kind of laboratory; in fact, it is the only place where the full range of selection pressures can be seen at work, including their numerous interactions." The studies of evolution and behavior may be joined in a common pursuit played directly in the field.

What to study? One need not travel to the rain forests of Brazil or those of the Congo to find suitable birds for study. Many of our most common species are not known in any detail, and the study of these would prove to be the most rewarding. Konrad Lorenz once said something to the effect that the answer to many of our problems in animal behavior may be swimming in the nearest roadside ditch. What about endangered species? The time may not be far off when man will have to make decisions about the wildlife that will share this plundered planet with him.

Knowledge of the behavior and ecology of such endangered species will make this grim task easier to make more wisely. Furthermore, detailed studies of endangered species may be all that will remain in the archives of the future, once the living forms are gone forever. Many such species, however, may be saved by rearing and keeping them in zoos. Although their natural habitats may be destroyed, knowledge of their requirements will enable us to adjust the strange environment of captivity to suit their needs. Studies made now may reap rich rewards in the future, but our species has a habit of letting such opportunities slip by.

What does all of this have to do with man directly? Is there such a thing as human ethology? Some ethologists think so, and a number of the more prominent ones have in fact shifted from studies of insects, birds and frogs to man. It is of interest to note how many ethologists began their careers by studying, say, the behavior of some fascinating species of butterfly or lizard, then later shifted to research on a mammal—perhaps a bear or a weasel—and finally went from a well-developed program of nonhuman primate research to man. One such researcher is Irenäus Eibl-Eibesfeldt, an Austrian ethologist who has made many contributions to this field. Eibl-Eibesfeldt has studied a host of organisms in detail, but he is now engaged in human ethological studies. He and one of his associates, H. Haas, make use of an ingenious device for recording observations of humans without their realization of it. If you aim the lens of a motion picture camera at a human, you are sure to get an unnatural record of the behavior on film. Eibl-Eibesfeldt and Haas make use of a camera that enables the photographer to aim away from the desired human subject, yet film him; this is done by the use of a prism built into a tube that fits over the end of the lens. The front of the lens is really a dummy; the actual record is taken through the prism on the side of the tube. The unsuspecting subject goes about behaving in a normal way in the belief that the photographer is filming something or someone in front of the camera. Such documentation of behavioral activities is just as vital in human ethology as it is in the study of bird behavior, but the researcher may have to be a bit more ingenious, as in the example just cited. It is quite clear that ethological studies of vanishing tribes and similar groups must be made before our opportunities for such research vanish along with the endangered species of the world.

In summary, there is much basic work that needs doing, and no time is better than the present to enter the mainstream of animal research, whether your subject be bird, bee or Bantu.

Additional Reading

Bastock, Margaret. 1967. Courtship: a zoological study. London: Heinemann Educational Books Ltd., 220 pp. (available in the U.S.A. from Aldine of Chicago). A very readable account of courtship and its physiological bases and evolutionary origins. Some basic biology at the college level would be helpful to fully profit from this technical account.

Beadle, George and Muriel. 1966. The language of life. Garden City, New York: Doubleday & Co., Inc., 242 pp. An extremely lucid account of heredity by a Nobel Laureate in Physiology and Medicine. First-rate for brushing up on your knowledge of genetics. Available in paperback.

Carthy, J. D. 1965. Animals and their ways; the science of animal behavior. Garden City, New York: Natural History Press, 156 pp. A very fine, easy-to-read-and-understand introduction to the world of behavior with numerous illustrations. Many bird examples.

Darling, Lois and Louis. 1968. A place in the sun; ecology and the living world. New York: William Morrow & Company, 128 pp. An elementary, clear account of some of the principles of ecology, illustrated by the authors.

Gilliard, E. Thomas. 1963. The evolution of bowerbirds. Scientific American (August), pp. 38–46. An interesting article in which Gilliard puts forth his idea that the evolution of bowerbird displays has shifted from the plumage of the birds to their external constructions—the decorated bowers built by these unusual birds.

Goodwin, Derek. 1965. Instructions to young ornithologists. VI. Domestic birds. London: Museum Press Ltd., 141 pp. Written for young adults, this is a very readable and informative account of man's bird domestications. The author is a famous authority on the pigeons of the world.

Howard, H. Eliot. 1964. Territory in bird life. New York: Atheneum, 239 pp. An inexpensive paperback reprint of a great classic in ornithology first published in 1920. Recent ideas and findings in this most important field of biology are given in the introduction to the reprint by Sir Julian Huxley and James Fisher.

Huxley, Julian S., and H. B. D. Kettlewell. 1965. Charles Darwin and his world. London: Thames and Hudson, 144 pp. A first-rate biography by two men who have contributed much to the evolutionary literature. Fine illustrations and easily understood text.

Johnsgard, Paul A. 1967. Dawn rendezvous on the lek. Natural History (March), pp. 16–21. These fascinating birds are here described in an easy-to-read popular account; illustrated.

Johnsgard, Paul A. 1968. The evolution of duck courtship. Natural History (February), pp. 4, 58–63. An interesting, illustrated account of the genetic background and evolution of duck displays by a leading student of the group.

Karp, Walter. 1968. Charles Darwin and the origin of species. New York: American Heritage Pub. Co., 153 pp. One of the Horizon Caravel Series, this is a nicely illustrated account of Darwin's life and scientific contributions. Easy to read.

Kruuk, Hans. 1966. Defense against killers. Natural History

(April), pp. 48–55. A popular account of how gulls protect themselves against enemies on the ground and in the air.

Lack, David. 1947. Darwin's finches. Cambridge, England: Cambridge University Press, 208 pp. A classic study of these now famous birds by one of the leading students of the group. This account is not easy reading, but is most rewarding and worth the effort expended.

Lack, David. 1953. Darwin's finches. Scientific American (April), pp. 16, 66–68. A much less technical account than Lack's 1947 book on the same subject. Line drawings of the finches illustrate this article.

Lack, David. 1965. The life of the robin. 4th edition. London: H. F. & G. Witherby, Ltd., 240 pp. A delightful book about the English Robin, not to be confused with the American Robin. Fine line drawings by Robert Gillmor. Territoriality, aggression, courtship, and other facets of behavior are treated in an easy style.

Lorenz, Konrad Z. 1958. The evolution of behavior. Scientific American (December), pp. 22, 67–74. The founder of modern comparative ethology uses a number of bird examples to discuss evolutionary pathways in behavior. Fine illustrations by Rudolf Freund, especially of duck behavior.

Maher, William J. 1966. Predation's impact on penguins. Natural History (January), pp. 42–51. A nice account of predator-prey relationships and their effects on the interacting species.

Meyerriecks, Andrew J. 1960. Success story of a pioneering bird. Natural History (August), pp. 46–57. A popular account of the spread of the Cattle Egret in the New World and some of the adaptations of this unusual member of the heron family.

Meyerriecks, Andrew J. 1961. Plumes, threats, and white beauty. Natural History (March), pp. 20–25. A popular account of heron courtship and related activities, with fine photographs by a leading Japanese photographer.

Meyerriecks, Andrew J. 1962. Diversity typifies heron feeding. Natural History (June), pp. 48–59. How do herons keep from competing with one another while foraging in the same area? The answer is contained in this popular account nicely illustrated by Guy Tudor.

Meyerriecks, Andrew J. 1962. Courtship in animals. BSCS Pamphlet Series, Number 3, 35 pp. (Biological Sciences Curriculum Study, Boulder, Colorado). A popular account of animal courtship behavior written for serious high school students. Many fine illustrations in color; a number of bird examples are used.

Peterson, Roger T., Guy Mountfort, and P. A. D. Hollom. 1966. A field guide to the birds of Britain and Europe. London: Collins, 344 pp. All of the British and European birds mentioned in the text can be found illustrated in this fine field guide.

Smith, Neal. 1967. Visual isolation in gulls. Scientific American (October) pp. 19, 94–102. A nicely illustrated account of how visual isolating mechanisms operate in a complex group of North American gulls.

Stebbins, G. Ledyard. 1966. Processes of organic evolution. Englewood Cliffs, New Jersey: Prentice-Hall, Inc., 191 pp. This is a recent, up-to-date coverage of the field of evolutionary biology for the college undergraduate. Many illustrations; a number of bird examples are used.

Thomson, A. Landsborough. 1964. A new dictionary of birds. New York: McGraw-Hill Pub. Co., 928 pp. This is a first-rate reference work for anyone interested in any aspect of bird biology. Many of the subjects covered in this book have review articles in the dictionary: evolution, lek birds, territory, brood parasitism, and so forth. A mine of information, technical but clearly written.

Thorpe, W. H. 1956. The language of birds. Scientific American (October), pp. 36, 128–30. An easy-to-read-and-understand introduction to bird song by one of the world's foremost researchers in this area.

Tinbergen, Nikolaas. 1960. The evolution of behavior in gulls. Scientific American (December), pp. 38, 118–26. The world's authority on gull behavior discusses in detail the communication systems used by gulls and how they got that way through time; nicely illustrated.

Tinbergen, Nikolaas. 1967. The Herring Gull's world; a study of the social behavior of birds. Garden City, New York: Doubleday & Co., Inc., 297 pp. A classic in the field of bird behavior. Professor Tinbergen has a clear, easy-to-understand style that is both entertaining and informative.

Tinbergen, Nikolaas. 1968. Curious naturalists. Garden City, New York: Doubleday & Co., Inc., 301 pp. A delightful account by the renowned naturalist, brimming with solid information gathered in the field by Professor Tinbergen and his many students. The author, Dutch by birth, now teaches at Oxford University in England.

Wallace, Bruce, and Adrian M. Srb. 1964. Adaptation. 2nd edition. Englewood Cliffs, New Jersey: Prentice-Hall, Inc., 115 pp. A good introduction to a variety of adaptations, their genetic basis, and how they originated and evolved. Many bird examples.

Index

[Page numbers in bold face refer to illustrations]